Deep Learning Enabled Semantic Communications

IEEE Press

445 Hoes Lane

Piscataway, NJ 08854

Deep Learning Enabled Semantic Communications

Zhijin Qin
Tsinghua University

Huiqiang Xie
Jinan University

Zhenzi Weng
Imperial College London

Xiaoming Tao
Tsinghua University

IEEE Press
WILEY

Published by John Wiley & Sons, Inc., Hoboken, New Jersey.

Library of Congress Cataloging-in-Publication Data has been applied for:

Hardback ISBN: 9781394306237
ePDF ISBN: 9781394306251
epub ISBN: 9781394306244
oBook ISBN: 9781394306268

Cover Design: Wiley
Cover Image: © ADDICTIVE STOCK/stock.adobe.com

Contents

Foreword

The rapid evolution of wireless communication technologies has consistently reshaped how we interact with information, devices, and ourselves. From the early days of voice calls to today's immersive 6G-enabled applications, such as real-time holographic communication, autonomous systems, and AI-driven services, the demand for intelligent, efficient, and context-aware networks has reached an unprecedented level. At the heart of this transformation lies the semantic communications, a paradigm shift poised to redefine the boundaries of connectivity by enabling machines not just to transmit data, but to understand and act on its meaning. As networks grow denser and applications become more sophisticated, traditional communication frameworks face unprecedented challenges on how to efficiently compress and transmit multimodal data, mitigate wireless distortions, and adapt dynamically to diverse tasks. To address these issues, the integration of deep learning into semantic communication offers a revolutionary approach to transform raw data into context-rich, task-oriented representations, so as to optimize bandwidth, enhance reliability, and unlock new possibilities for intelligent systems.

This book, entitled *Deep Learning Enabled Semantic Communications*, offers a timely and comprehensive exploration of this cutting-edge field. Structured systematically, it begins with foundational concepts of deep learning and semantic information theory, and then progresses to advanced architectures and robust designs to transmit multimodal data spanning text, speech, image, and video. What makes this book particularly compelling is its coverage of a rapidly evolving field, its systematic organization, and the depth of insight it provides into both fundamental theories and practical implementations. By bridging deep learning with semantic communications, it delivers a clear, authoritative, and forward-looking exploration of how information can be transmitted and interpreted more intelligently. This comprehensive treatment makes it a seminal book for current applications and a foundation for future innovations in communication technology.

This book will serve as an essential reference for engineers and industry practitioners, as well as researchers, providing up-to-date coverage of semantic communications. Additionally, this classic text will contribute to laying the groundwork for transformative applications of semantic communications that could impact industries and society globally.

Professor Zhu Han
Fellow of IEEE/ACM/AAAS
University of Houston

Preface

The great potential of deep learning in addressing challenging problems of wireless communications has been increasingly acknowledged by domain experts, and a significant amount of research has since been dedicated to this emerging area, whose reach and impact keep growing. In recent years, semantic communications research has become widely popular in academia and industry. Its deployment in the sixth generation (6G), where intelligence is expected to permeate, has directly motivated the writing of this book. We believe this is a good time to publish an informative yet accessible book by a credible publisher that systematically documents these exciting scientific advances and, more importantly, helps readers thread through the morass of immense recent research findings.

It is for this purpose that this book provides a detailed treatment of deep learning enabled semantic communication systems. It has a concise and focused structure that threads and positions significant concepts and methods to form a landscape, making it much easier for readers to gain a deep and insightful understanding of the subject. Thanks to the short and sweet exposition in the book, the readers are expected to appreciate the benefits of the interplay between wireless and deep learning and continue contributing to the booming research activities and mobilizing the methods and concepts in practical system design. We believe this book will become the go-to resource for researchers and practitioners in wireless communications.

To achieve these objectives, this book provides a focused introduction to the interplay between semantic communications and deep learning. Deep learning has increasingly become a popular and effective research field. Particularly, the book extensively discusses the use of deep learning to revolutionize semantic communications and develop task-oriented semantic transmission paradigms in wireless communication systems. Furthermore, it explores how deep learning can be utilized to analyze and perceive semantic information, and further enable interactive semantic communications.

Organization of the Book

Chapter 1 includes a comprehensive introduction to Shannon's three-level semantic communication theory, followed by a detailed discussion of the fundamentals of semantic communications and cutting-edge deep learning algorithms before transitioning into the theoretical principles and practical significance of deep learning enabled semantic communications. Besides, it elaborates on the essential role of semantic communications in the intersection of 6G.

Chapter 2 begins with the background of semantic information theory. Firstly, this chapter introduces the different definitions of semantic entropy. Besides, semantic mutual information and the semantic channel capacity will be illustrated. In addition, the semantic source coding and channel coding theorem are reviewed. Finally, the information bottleneck is discussed.

Chapter 3 explores semantic communication systems for source reconstruction. It encompasses the design of a unique neural network for extracting and representing semantic information and facilitating efficient transmission and recovery. Furthermore, this chapter introduces performance metrics for measuring semantic errors, which helps optimize the system at the semantic level.

Chapter 4 presents the motivations behind task-oriented semantic communication systems and elaborates on the technical problems faced by task-oriented semantic communication systems. Then, it presents the framework of task-oriented semantic communication systems and the functionalities implemented by each module of the framework. Finally, this chapter describes various intelligent tasks according to different sources.

Chapter 5 exploits the influence of semantic information on the optimal data sampling and compressing strategy. It improves the sensing efficiency by adjusting the measurement number and biases in accordance with the semantic information in the source. This chapter further explores the joint design of data sampling, compression, and coding schemes under the guidance of semantic information.

In Chapter 6, the robust semantic communication systems that strive to minimize semantic impairments for achieving accurate semantic transmission are introduced. In particular, we provide an overview of various forms of semantic impairments, such as literature semantic noise and adversarial semantic noise. Then, the robust system model is presented, which allows for effective semantic exchange in the presence of different forms of semantic impairments.

Chapter 7 explores various applications of generative AI in semantic communications across different domains. It begins with immersive communications, highlighting how GAI enables the creation of virtual environments, avatars, and interactive narratives. It then shifts to autonomous driving, where GAI processes sensor data to improve the understanding of a vehicle and its surroundings. In the context of smart cities and satellite communications, GAI analyzes data streams like traffic flow and energy usage to optimize city operations and improve public safety.

Chapter 8 provides a brief summary of the deep learning enabled semantic communication techniques discussed in this book. It discusses some research directions that are of great significance and need to be addressed in future semantic communications, including one model for all semantic communications, semantic communications compatibility, semantic communications standardization, and neuro-symbolic AI-enabled semantic communications.

Acknowledgments

We would like to thank all the individuals and institutions who have contributed to this book. This work is the result of collaborative efforts from researchers across multiple disciplines, and it would not have been possible without the support, encouragement, and insights of many.[1]

We would especially like to acknowledge the contributions of Xiang Peng, Wanli Ni, Yinhuan Huang, and Zhiyuan Qi from Tsinghua University, Beijing, China, for their valuable contributions in reviewing the manuscript and offering constructive feedback during the writing process. Their discussions, suggestions, and shared expertise have greatly enriched our understanding of the semantic communications field and helped improve the quality of this book.

Additionally, we acknowledge the contributions of the broader research community, whose pioneering works in deep learning, information theory, and communication systems have laid the foundation for the advancements presented in this book. The open-source initiatives and publicly available datasets have particularly benefited our experimental studies and analyses.

Finally, we are grateful to our publishers and editorial team for their professional guidance and assistance in bringing this book to fruition. Their dedication and attention to detail have greatly enhanced the quality of this work. We hope that this book serves as a valuable resource for researchers, practitioners, and students interested in the future of semantic communications.

Z. Q.

1 This work was supported in part by the National Key Research and Development Program of China under Grant 2023YFB2904300; in part by the National Natural Science Foundation of China (NSFC) under Grant 62293484 and Grant 62401227; in part by Guangdong Basic and Applied Basic Research Foundation under Grant 2025A1515010104.

Acronyms

AI	Artificial Intelligence
AoI	Age of Information
AR	Augmented Reality
ASR	Automatic Speech Recognition
AWGN	Additive White Gaussian Noise
BER	Bit-Error Rate
BERT	Bidirectional Encoder Representations from Transformers
Bi-LSTM	Bidirectional Long Short-Term Memory
BLEU	Bilingual Evaluation Understudy
BPG	Better Portable Graphics
CE	Cross-Entropy
CNN	Convolutional Neural Network
CSI	Channel State Information
CV	Computer Vision
dB	Decibel
DL	Deep Learning
DNN	Deep Neural Network
E2E	End-to-End
FFN	Feed-Forward Network
FGM	Fast Gradient Method

GAN	Generative Adversarial Network
GPT	Generative Pre-Trained Transformer
GRU	Gated Recurrent Unit
IoT	Internet-of-Things
JSCC	Joint Source-Channel Coding
KPI	Key Performance Indicator
LDPC	Low-Density Parity-Check Codes
LPIPS	Learned Perceptual Image Patch Similarity
LS	Least Squares
LSTM	Long Short-Term Memory
MDC	Multiple Description Coding
MHSA	Multi-Headed Self-Attention
MIMO	Multiple-Input Multiple-Output
ML	Machine Learning
MOS	Mean Opinion Score
MSE	Mean-Squared Error
NLP	Natural Language Processing
NN	Neural Network
PESQ	Perceptual Evaluation of Speech Quality
PSNR	Peak Signal-to-Noise Ratio
QoE	Quality of Experience
QPSK	Quadrature Phase Shift Keying
ReLU	Rectified Linear Units
RNN	Recurrent Neural Network
SER	Symbol-Error Rate
SISO	Single-Input Single-Output
SKB	Semantic Knowledge Base
SNR	Signal-to-Noise Ratio
STS	Semantic Textual Similarity
SVM	Support Vector Machine
VAE	Variational Autoencoder
VoI	Value of Information

VQA	Visual Question Answering
VR	Virtual Reality
WER	Word Error Rate
5G	Fifth Generation
6G	Sixth Generation

Notation

$\triangleq$	Defined as equal to ($a \triangleq b$: a is defined as b)
$\approx$	Approximately equal to
$\propto$	Proportional to
$\odot$	Hadamard product (entry-wise multiplication)
$\otimes$	Kronecker product operator
$*$	Circular convolution operator
$f : \mathcal{X} \mapsto \mathcal{Y}$	Function f from elements of set $\mathcal{X}$ to elements of set $\mathcal{Y}$
$\leftarrow$	Assignment
$\mathbb{R}$	Field of all real numbers
$\mathbb{R}^+$	Field of all positive real numbers
$\mathbb{C}$	Field of all complex numbers
$\mathbf{1}\{predict\}$	Indicator function ($\mathbf{1}\{predict\} = 1$ if the predict is true, else 0)
$\boldsymbol{A}^H$	Hermitian (conjugate transpose) of matrix $\boldsymbol{A}$
$\boldsymbol{A}^T$	Transpose of matrix $\boldsymbol{A}$
$\boldsymbol{A}^{-1}$	Inverse of matrix $\boldsymbol{A}$
$\|\|\boldsymbol{A}\|\|_F$	Frobenius norm of matrix $\boldsymbol{A}$
$\det(\boldsymbol{A})$	Determinant of matrix $\boldsymbol{A}$
$\mathrm{tr}(\boldsymbol{A})$	Trace of matrix $\boldsymbol{A}$
$\mathrm{diag}([x_1, \ldots, x_N])$	The $N \times N$ diagonal matrix with diagonal elements $x_1, \ldots, x_N$

$\|\|\boldsymbol{x}\|\|$	Norm of vector $\boldsymbol{x}$
$\boldsymbol{x}^*$	Complex conjugate of vector $\boldsymbol{x}$
$\boldsymbol{x}^H$	Hermitian (conjugate transpose) of vector $\boldsymbol{x}$
$\boldsymbol{x}^T$	Transpose of vector $\boldsymbol{x}$
$\mathbf{0}$	The zero vector
$\mathbf{I}$	The identity matrix
$\mathbb{E}$	Expectation of a random variable X
$\mathrm{Cov}[X, Y]$	Covariance of random variables X and Y
$Pr\{X = x\}$	Probability that a random variable X takes on the value x
$X \sim p_X(x)$	The random variable X has distribution $p_X(x)$
$\mathcal{N}(\mu, \sigma^2)$	Real Gaussian (normal) distribution with mean μ and variance σ^2
$\mathcal{CN}(\mu, \sigma^2)$	Circularly symmetric complex Gaussian (normal) distribution with mean μ and variance σ^2
$\arg\max f(x)$	Value of x that maximizes the function $f(x)$
$\arg\min f(x)$	Value of x that minimizes the function $f(x)$
$\mathrm{Re}\{x\}$	Real part of x
$\mathrm{Im}\{x\}$	Imaginary part of x
$\|x\|$	Absolute value (amplitude) of x
$\frac{\partial y}{\partial x}$	Partial derivative of y with respect to x

1

Introduction

Wireless communication systems have undergone vigorous advancements from the first generation (1G) to the fifth generation (5G) over the past few decades by developing numerous coding algorithms and channel models to recover accurate sources at the bit level. However, in recent years, the flourishing of artificial intelligence (AI) has revolutionized various industries and incubated multifarious intelligent tasks, which increases the amount of data transmission to the zetta-byte level and requires massive machine connectivity with low transmission latency and energy consumption. In this context, conventional communication systems face severe challenges imposed by ubiquitous AI tasks. Therefore, it is inevitable that a new communication paradigm needs to be developed. Semantic communications have been proposed to address the challenges by extracting semantic information inherent in source data while omitting irrelative redundant information to reduce the transmission data, thereby lowering communication resources and facilitating high semantic fidelity transmission. Nevertheless, the exploration of semantic communications has gone through decades of stagnation since it was first identified because of the inadequacy of mathematical models for semantic information. Inspired by the thriving of AI, deep learning (DL)-enabled semantic communications have been scrutinized as promising solutions to the bottlenecks in conventional communications by leveraging the learning and fitting capabilities of neural networks (NNs) to bypass mathematical models for semantic extraction and representation.

To this end, this book introduces DL-enabled semantic communications, including concepts, applications, and challenges. In particular, it comprehensively covers semantic communications for source reconstruction, task-oriented semantic communications, semantic impairments, semantic knowledge bases (SKBs), and large model-driven semantic communications. The objective of this book is

Deep Learning Enabled Semantic Communications, First Edition. Zhijin Qin et al.

Published 2026 by John Wiley & Sons, Inc.

to help readers gain a fundamental appreciation and understanding of the emerging DL-enabled semantic communications and their implications for future intelligent wireless networks.

1.1 Conventional Communications versus Semantic Communications

In this section, we introduce the basic yet important theory of wireless communications and briefly present the history of semantic communications.

1.1.1 Three-level Communications

In 1949, Shannon and Weaver [1] categorized communications into three levels:

- *Level A*: how accurately can the symbols of communication be transmitted? (The technical problem)
- *Level B*: how precisely do the transmitted symbols convey the desired meaning? (The semantic problem)
- *Level C*: how effectively does the received meaning affect conduct in the desired way? (The effectiveness problem)

The first level of communications is to deliver the symbol transmission accurately, that is, syntactic communications, which improves the data rate by expanding the bandwidth resources, increasing the transmission power, and adding transmission antennas. The 1G to 5G wireless communication networks belong to syntactic communications, and the system capacity has been proven to be approaching the Shannon limit by utilizing efficient source coding and channel coding algorithms [2]. The different coding modules in conventional communication systems are designed and optimized separately, which converts the input message into the bit sequence and focuses on performance improvement at the bit or symbol level by taking bit-error rate (BER) or symbol-error rate (SER) as metrics to measure the information loss. However, the bit-oriented transmission framework requires the precise alignment of input and recovery messages. Still, it neglects the underlying meaning behind bits, which augments the possibility of transmitting unnecessary data beyond user requirements, accelerating the consumption of communication resources and increasing transmission latency. Moreover, due to the wide deployment of Internet-of-things (IoT) applications, conventional communications are no longer ideal as they transmit information that could be irrelevant to the downstream intelligent tasks at the receiver.

To this end, the second level of communications, that is, semantic communications, considers the inherent semantics of input information to tackle the technical problems in the bit-oriented paradigm. Semantics take into account the meaning and veracity of source information because they can be both informative and factual [3]. Besides, semantic data can be compressed to a proper size for transmission by using a lossless method [4], which leverages the semantic relationship between different messages, while the traditional lossless source coding represents a signal with the minimum number of binary bits by exploring the dependencies or statistical properties of the input message. Moreover, the semantic information varies for different transmission purposes, which could be in various formats, for example, age of information [5], or more complicated semantic features. Inspired by this, semantic communications deviate from the bit-oriented paradigm by extracting semantics from the input message with minimal ambiguity to facilitate semantic exchange between the transmitter and the receiver, committing to reducing network traffic by transmitting low-dimensional semantic information and optimizing the system performance by minimizing the semantic error instead of the BER or SER. It is worth mentioning that there is currently no consensus on the definition of semantic error in the field of semantic communications. In some works, semantic errors refer to inaccurate semantics that lead to misunderstanding and ambiguity, such as spelling errors in the text.

Furthermore, the third level of communications involves the effectiveness of semantic transmission, that is, pragmatic communications, which condenses the input message to obtain the semantics only associated with user requests. Inspired by this mechanism, goal/task-oriented semantic communications are boosted to perform specific downstream tasks required by receiver users by delivering task-related semantic features [6]. Conventional communication systems are constrained to achieve source reconstruction in the same modal before performing any modal conversion at the receiver. However, in goal/task-oriented semantic communications, the recovered message is no longer confined to the same modality as the input message, facilitating flexible cross-modal or single-modal to multimodal transmissions, such as speech-to-text and text-to-image/speech, to satisfy different user requests and improve the user quality of experience (QoE).

1.1.2 History of Semantic Communications

Decades ago, the concept of semantic communication was pioneered by engineers and philosophers. As far back as 1925, Dewey [7] asserted that "communication should be regarded as a mechanism for attaining purposes," while Wittgenstein [8] emphasized the centrality of meaning in philosophical discourse. Later, in 1938, Cherry went on to define semantics in terms of signs, which fall into the following categories:

- Syntax: studies the signs and their relations to other signs
- Semantics: studies the signs and their relation to the world
- Pragmatics: studies the signs and their relations to users

Potentially influenced by the aforementioned definition, in 1949, Weaver [1] formulated semantics with the perspective of engineering and outlined three communication levels, that is, *technical-, semantic-, and effectiveness-levels*, as noted earlier. Shannon defined communication as the precise or approximate reconstruction of information from one point to another, falling under the technical level. Consequently, Shannon applied statistical methods to quantify the uncertainty of information while overlooking the semantic and effectiveness levels. Despite Shannon's clear indication, many researchers still tried to adapt the statistical framework to interpret or evaluate semantics. Carnap and Bar-Hillel [3] first attempted to develop a "theory of semantic communication" in 1952 with the concept of logical probability. Since then, explorations into semantic communication have continued steadily.

Up until recently, semantic communication gained more and more attention as a possible solution to enable more efficient ways to exchange information. In 2003, Jeong et al. [9] first employed a knowledge graph in semantic communications. After one year, Floridi et al. [10] outlined the "theory of strongly semantic information." In 2011, Bao et al. [11] proposed a generic model of semantic communications. After a period in the wilderness, the bloom of DL has prepared the way for all progress and development of semantic communications because of the significant progress in semantic extraction and understanding. For example, the remarkable works [12, 13] designed the initial deep joint source-channel coding (JSCC) for text transmission and proposed a single-user joint semantic-channel coding in 2018 and 2021, respectively. After that, the research of semantic communications mainly focuses on the DL-enabled paradigm.

1.2 Introducing Deep Learning to Semantic Communications

We first introduce the concept of DL and present some cutting-edge algorithms relevant to the following chapters. Then, we discuss the motivations and details of applying deep learning in semantic communication systems.

1.2.1 Deep Learning Basics

DL is defined as a subset of machine learning (ML) that falls under the umbrella of AI. DL leverages numerous classifiers working together based on linear regression followed by specific activation functions. The principle behind DL mirrors that of conventional statistical linear regression, albeit with considerable neural nodes in DL instead of just one node in traditional approaches. These neural nodes are connected through complicated topological structures to form a NN. The motivation of the NN derives from the human brain, which allows it to mimic intricate mathematical patterns and learn data representations through hierarchical intelligence.

Inspired by this, DL leads to unprecedented success in various areas, such as computer vision (CV) and natural language processing (NLP). The fundamental DL algorithms are briefly introduced in the following sections.

1.2.1.1 Fully Connected Neural Network

A fully connected NN is a classic DL model consisting of multiple fully connected layers, also termed dense layers, where each neuron establishes connections with all neurons in the preceding layer. Based on sufficient training data, the fully connected NN adaptively adjusts its weights and/or bias matrices to steer the output toward the predetermined target. An example of the fully connected NN is shown in Figure 1.1. From the figure, the output of each neuron can be denoted as:

$$\text{Output}_j = f_a\left(\sum_{i=1}^{I} (\mathcal{W}_{i,\,j} \times \text{Input}_i) + bias_j\right), \tag{1.1}$$

where $\mathcal{W}_{i,\,j}$ represents the trainable weight corresponding to the i-th neuron in the previous layer to the j-th neuron in the current layer; bia_j and $f_a(\cdot)$ denote the bias of the j-th neuron in the current layer and the adopted activation function, respectively.

The fully connected NN has proved remarkable adaptability and compatibility with multimodal data and intricate cognitive tasks, owing to its robust nonlinear learning and fitting prowess. Nevertheless, it is susceptible to overfitting and demands tremendous computational resources for efficient model training. Consequently, it usually synergizes with other DL algorithms to construct a sophisticated NN architecture.

1.2.1.2 Convolutional Neural Network

The inception of the convolutional neural network (CNN) can be attributed to LeCun et al. [14], who initially proposed it as an end-to-end (E2E) image analysis framework tailored for handwritten digit recognition. Over the intervening

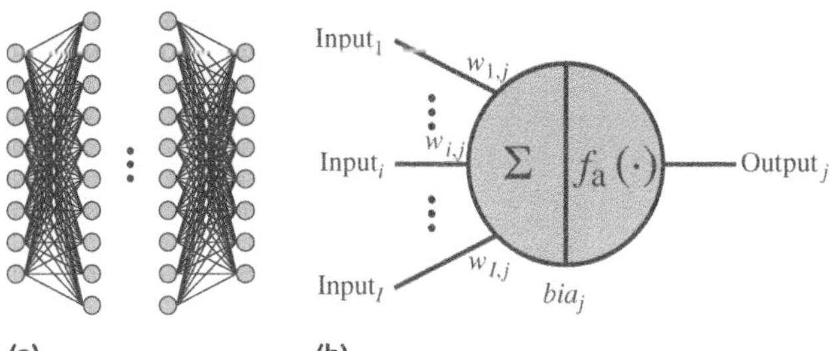

Figure 1.1 (a) The fully connected NN architecture; (b) one neuron processing unit.

decades, CNN and its variants have shown significant superiority in handling visual data, surmounting substantial bottlenecks in CV that exceed human limits. The central concept of CNN is the comprehensive processing and fusion of feature maps, enabling the inference of nonlinear correlations between input data and established ground truths.

The typical CNN architecture for the image classification task is shown in Figure 1.2. From the figure, the CNN architecture consists of the upstream feature extractor and the downstream classifier. The backbone of the feature extractor includes multiple convolutional layers and pooling layers. In particular, the convolutional layer meticulously orchestrates the output of neurons connected to local regions of the input image by utilizing the property of translation invariance and determines the output of each neuron by calculating scalar products between neural weights within the convolutional kernel and the corresponding input region. Subsequently, the pooling layer adeptly executes the spatial downsampling on the output of the convolutional layer to diminish the number of parameters during the activation process. Note that the depth of extracted features depends on the number of kernels, and the trainable weights within these kernels are updated during the training phase. Moreover, the classifier contains tandem fully connected layers, transforming the extracted features into class predictions and returning a probability vector. As a result, the accurate classification label for the input image can be inferred by indexing the position within the predefined label vector that corresponds to the highest probability.

CNNs exhibit lower complexity without sacrificing performance compared to fully connected NNs and dramatically mitigate the risk of overfitting through sharing parameters across the network. In addition, CNN is designed based on parallel computing architecture, providing great superiority for acceleration when implemented on hardware. CNNs are considered a pivotal DL algorithm and have been

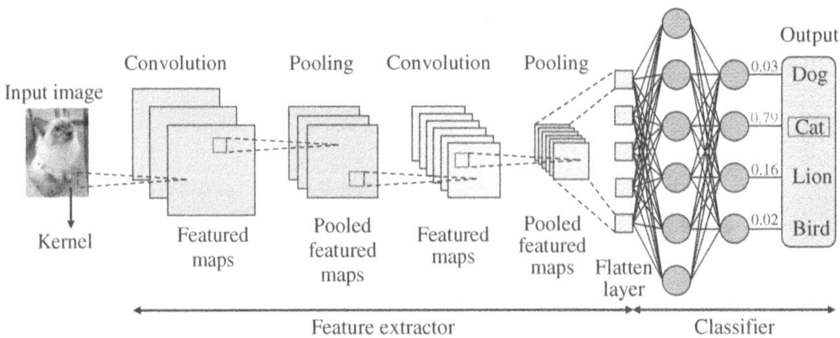

Figure 1.2 The typical CNN architecture for image classification.

extended from image processing to multimodal, for example, text and speech. Furthermore, the compatibility and universality of CNNs have been verified across diverse domains.

1.2.1.3 Recurrent Neural Network

The Hopfield network [15] developed the precursor of the recurrent neural network (RNN), showcasing the feasibility of recurrent structures within NNs. RNNs are bespoke to detect patterns in sequential data by maintaining an internal state, that is, memory, that captures information about previous inputs, allowing it to demonstrate dynamic temporal behavior and perform intelligent tasks, for example, time series prediction, machine translation, and speech recognition. RNNs are structured with recurrent connections, facilitating the effective extraction of features and modeling of dependencies inherent in sequential data. The basic RNN architecture is shown in Figure 1.3. From the figure, the input from the previous layer at the current time step and the hidden state of the current layer at the previous time step are both fed into the hidden state of the current layer at the current time step to produce the output.

However, the phenomenon of memory decay amplifies with the continuous expansion of temporal sequences, constraining the capacity to understand long-term dependencies and precipitating the predicament of gradient vanishing/-exploding. Toward this end, variants of the standard RNN, for example, long short-term memory (LSTM) [16] and gated recurrent unit (GRU) [17], were proposed to contend with the challenge of prolonged temporal interdependencies by incorporating specialized mechanisms to better process and preserve information over time.

The structure diagrams of the LSTM unit and the GRU unit are shown in Figure 1.4. From the figure, the LSTM unit introduces an innovative module, the memory cell, to control the information flow and maintain the ability to retain long-term memory. Specifically, the memory cell consists of the forget gate,

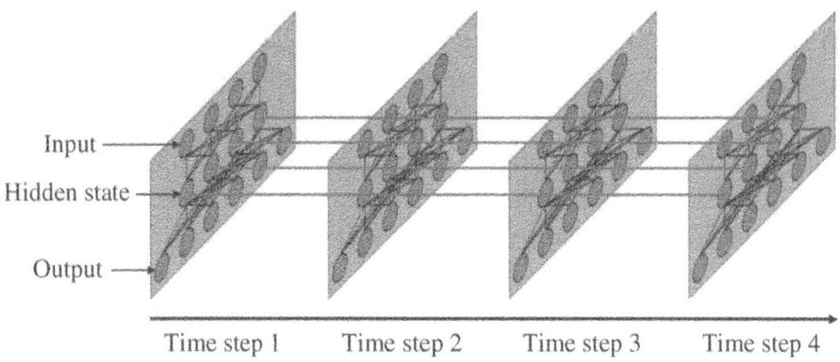

Figure 1.3 The basic RNN architecture.

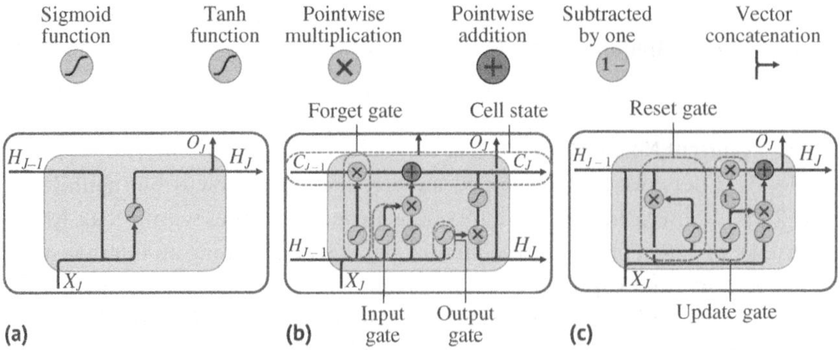

Figure 1.4 The structure diagrams of (a) RNN unit, (b) LSTM unit, (c) GRU unit.

the input gate, and the output gate, taking responsibility for adjudicating which information should be forgotten, managing the input of new information, and supervising the generation of the hidden state, respectively. Inspired by the LSTM unit, the GRU unit modifies the gate-controlled architecture and simplifies the data flow process by omitting the memory cell, thereby alleviating the computing complexity. Moreover, the reset gate and the update gate are devised to enable the reset of information from the previous time step and regulate the degree of integration of new information.

RNNs and RNN variants, for example, bidirectional RNN (BRNN) [18], have demonstrated superior performance compared to CNNs in intelligent tasks related to time series analysis. Nonetheless, the limited attention of RNNs restricts the comprehension of vast multilingual information, while the training complexity and instability hinder the deployment of RNNs in large language models.

1.2.1.4 Transformer

The transformer model [19] was originally carried out to implement machine translation tasks, relying solely on the novel fully connected layer-enabled attention mechanism and completely dispensing with RNNs and CNNs. The basic transformer architecture is shown in Figure 1.5. From the figure, the transformer skeleton comprises multiple encoder blocks and decoder blocks. The positional encoding is injected into the embedded tokens to record the relative or absolute positions of tokens within the sequence. At the encoder, the multi-headed self-attention (MHSA) module is invoked to facilitate comprehensive token representation learning while simultaneously identifying critical tokens. Besides, it leverages a softmax activation to calculate the self-attention based on three projections of tokens, that is, *query*, *key*, and *value*, denoted as

$$\text{Attention}(\mathcal{Q}, \mathcal{K}, \mathcal{V}) = \text{softmax}\left(\frac{\mathcal{Q}\mathcal{K}^T}{\sqrt{d_k}}\right)\mathcal{V}, \tag{1.2}$$

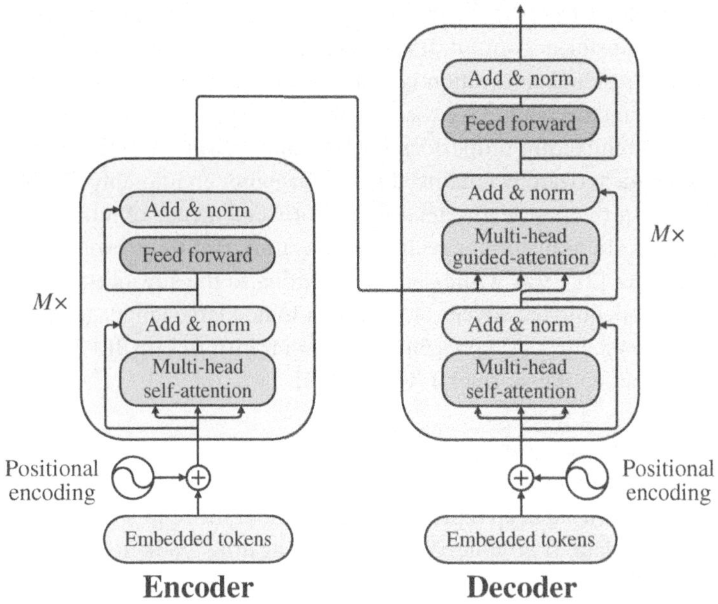

Figure 1.5 The basic transformer architecture.

where $\mathcal{Q}$, $\mathcal{K}$, and $\mathcal{V}$ represent projections *query*, *key*, and *value*, respectively. The scaling factor, $\sqrt{d_k}$, is the dimension of the fully connected layer in the *key* projection.

The feed forward network (FFN) reduces the intricacy of the MHSA output and extracts local patterns within the input sequence by introducing nonlinear transformation, which supports varied attention allocation and independent processing of each position. In FNN, the rectified linear units (ReLU) activation function is adopted between two fully connected layers. Therefore, the processing of the MHSA output, $\mathcal{O}_{\text{MHSA}}$, can be expressed as

$$\text{FFN}(\mathcal{O}_{\text{MHSA}}) = \max(0, \mathcal{O}_{\text{MHSA}}\mathcal{W}_1 + \boldsymbol{bias}_1)\mathcal{W}_2 + \boldsymbol{bias}_2, \tag{1.3}$$

where $\mathcal{W}_1$ and $\mathcal{W}_2$ are the weight matrices of the first and second fully connected layers, and $bias_1$ and $bias_2$ are the corresponding bias vectors, respectively.

Similarly, the decoder block involves three components, including the MHSA module, the multi-head guided-attention module, and the FNN. Specifically, the embedded tokens at the decoder are fed to the MHSA module after the positional encoding and produce the output as *query* position for the multi-head guided-attention module. In addition, the FNN output from the encoder is passed to the decoder and adopted as the input positions, *key* and *value*, for the multi-head guided-attention module. The FNN further refines the learned token representation and attains the transformer output. It is noteworthy that the layer normalization followed by the residual connection is employed for all modules across the transformer encoder and decoder.

The transformer architecture outperforms RNNs and CNNs on multifarious intelligent tasks, characterized by enhanced parallelizability and notably diminished training time. In the past few years, the meteoric emergence of numerous foundation models in the transformer realm, such as universal transformer [20] and vision transformer [21], was witnessed, contributing to the advancement of the NLP field and propelling the advent of the much-touted large language models, such as bidirectional encoder representations from transformers (BERT) [22] and generative pre-trained transformer 4 (GPT-4) [23].

1.2.1.5 Generative Adversarial Network

The generative adversarial network (GAN) model is an emerging technique that offers a solution for acquiring deep representations without the requisite of enormous labeled training data. It is achieved by backpropagating gradients through a pair of competing NNs to estimate the high-dimensional data distribution. The GAN framework was first proposed to generate realistic images [24], and the model structure is shown in Figure 1.6. From the figure, the GAN model includes two major parts: the generator and the discriminator. The generator has no access to real images but functions as a forger to synthesize plausible data from the latent space input. The discriminator plays the role of an expert in distinguishing between the real labels from the training set and the fake labels generated by the generator, providing a simple ground truth of the discriminative result. The ultimate goal of the generator is to create forgeries indistinguishable from the real data to deceive the discriminator. In contrast, the discriminator continuously sharpens its observation to detect fake samples, driving both entities to evolve and reinforce competitive capabilities to combat each other.

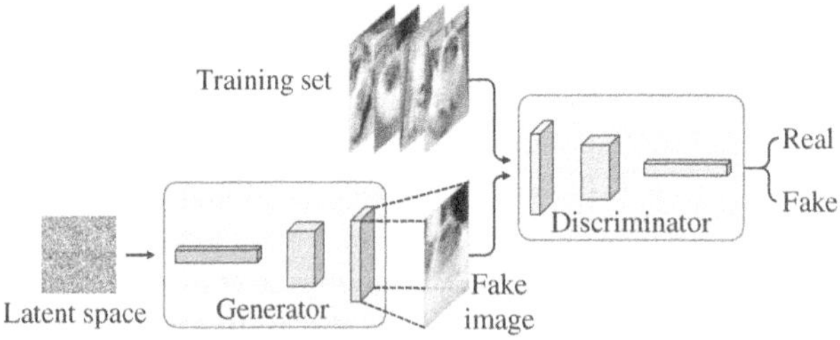

Figure 1.6 The GAN model structure for image generation.

The generator and discriminator are typically implemented using multiple NNs, such as CNNs and RNNs, for various generation tasks, adopting an alternate training manner to ensure iterative weight updates. The loss function to train the discriminator can be denoted as

$$\mathcal{L}_{\mathrm{D}} = \frac{1}{2} \log \mathfrak{T}_{\mathrm{D}} (S) + \frac{1}{2} \log (1 - \mathfrak{T}_{\mathrm{D}} (\mathfrak{T}_{\mathrm{G}} (\boldsymbol{Z}))), \tag{1.4}$$

where $\mathfrak{T}_{\mathrm{D}}(\cdot)$ and $\mathfrak{T}_{\mathrm{G}}(\cdot)$ are the discriminator and the generator, respectively. S is the real data, and $\boldsymbol{Z}$ is the latent space. Note that the NN parameters of the generators remain fixed throughout the discriminator's entire training process. Besides, the loss function can be varied according to different tasks.

Similarly, the loss function to train the generator can be expressed as

$$\mathcal{L}_{\mathrm{G}} = \log (1 - \mathfrak{T}_{\mathrm{D}} (\mathfrak{T}_{\mathrm{G}} (\boldsymbol{Z}))), \tag{1.5}$$

where the loss convergence is stabilized until the discriminator can no longer detect counterfeits of the generator from the real data. The NN parameters of the discriminator are fixed when training the generator.

GANs have experienced swift development over the past decade, and many variants have been developed, including conditional GANs [25] that control the generation by introducing additional information and deep convolutional GANs [26] that tackle the inherent training problems. However, further endeavors are necessary to tackle many challenges in GANs, such as the mode collapse, the training instability, and the ethical considerations regarding the generated content.

1.2.2 Deep Learning Enabled Semantic Communications

The conventional bit-oriented transmission protocol has evolved from syntax communications and undergone thriving developments from the 1G to the 5G wireless communication networks. Nonetheless, attention to semantic information has been neglected in conventional communications and ubiquitous intelligent tasks in the 5G and beyond communication era have aggravated the burden of data transmission to satisfy personalized and multifarious user requests. In this context, the semantic communication problem has been listed as one of the promising techniques for future intelligent communications to alleviate the scarcity of bandwidth resources and the explosion of data traffic [27–29]. Therefore, it is imperative to reconceive the framework of semantic communications and establish an innovative architecture capable of catering to escalating transmission requirements and serving on-demand mobile applications.

Thanks to the success of AI in recent years, a DL-enabled semantic communication paradigm has been proposed and has shown its great potential to tackle the technical challenges across many aspects of conventional communications, for example, physical-layer communications [30, 31] and wireless resource allocation [32, 33]. Such success is attributed to the learning and fitting capabilities of

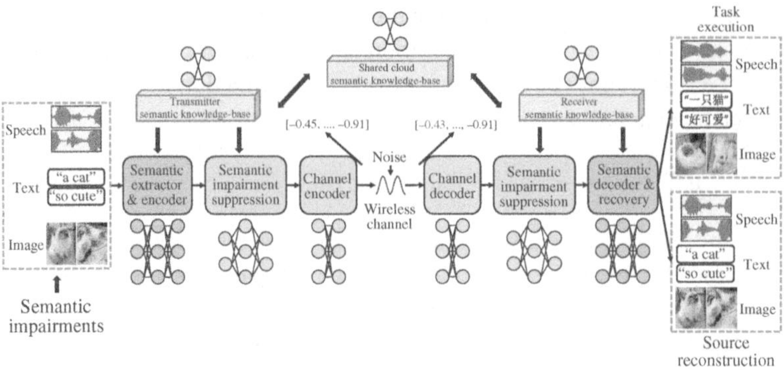

Figure 1.7 The structure of the DL-enabled semantic communication framework.

sophisticated NNs, breaking the constraint of a mathematical model to represent and exchange semantic information. A general DL-enabled semantic communication system is shown in Figure 1.7. From the figure, the semantic encoder takes responsibility for extracting the semantic information from the input source, and the semantic impairment suppression mechanism commits to perceiving and alleviating semantic impairments. The transmitter and receiver SKBs are invoked to better facilitate semantic representation and semantic recovery, respectively. The DL-enabled semantic communication paradigm requires the least semantic ambiguity between the source and the recovered messages. It maps the source message into low-dimensional semantic features, reducing the volume of transmission data and mitigating the network traffic without performance degradation. According to the cutting-edge DL-enabled semantic communications, the transmission goal could be categorized into two types: source data reconstruction and intelligent task execution. It is noteworthy that the DL-based JSCC also adopts NNs to replace conventional coding and decoding modules. The principal differences are clarified as follows to clarify the distinction between DL-based JSCC and DL-enabled semantic communications.

- The DL-based JSCC is a semantic-agnostic solution that jointly optimizes the source compression and error correction by training an integrated NN to improve the E2E system performance. Meanwhile, DL-enabled semantic communications utilize NNs to learn critical semantic information from the source and facilitate semantic transmission to convey the meaning of the transmitter to the receiver.
- The JSCC architecture only considers the physical noise, whereas semantic communications take into account both the physical noise and the specific semantic impairments. Consequently, DL-enabled semantic communications incorporate the semantic impairment suppression module to mitigate the effects of semantic impairments.

- JSCC has no SKBs. DL-enabled semantic communications synergize with SKBs to ensure high semantic fidelity transmission.

SKBs are memorable knowledge network models that are intricately structured and offer relevant semantic descriptions for data. As shown in Figure 1.7, SKBs are divided into the transmitter SKB, the shared cloud SKB, and the receiver SKB. In semantic communications, the semantic encoder produces the compressed codestream based on the learning and inference algorithm, where the transmitter SKB acts as a booster to accelerate the learning process and standardize the optimal inference path by providing multi-level semantic knowledge vectors. The description of propagation environments is shared between the transmitter and the receiver based on the shared cloud SKB to facilitate high semantic fidelity transmission. At the receiver, the resilient decoding mechanism is adopted to calibrate corrupted semantic features by sequentially incorporating received semantic knowledge vectors from the highest level to the lowest level.

DL-enabled semantic communications for source reconstruction have proven superior in tackling the performance bottlenecks of conventional communications. To achieve precise source reconstruction, the global semantic information is extracted for transmission. To serve intelligent downstream tasks, only task-related semantic features are extracted, while other irrelative features are ignored to minimize the transmitted data. Besides, the produced data is not required to exhibit precise alignment with the source data at the binary level, and it typically diverges from the source. More details of DL-enabled semantic communications and task-oriented semantic communications are introduced in Chapters 3 and 4, respectively.

1.3 Semantic Communications for Further Networks

The previous 5G has rapidly changed the lifestyle of humans and given birth to connected intelligence, which transforms society and industry into data-centric and automated. In the era of 6G and beyond, it is anticipated that the trend will be further enhanced. The fusion of digital and physical realities, combined with intelligently interconnected systems, is fueling the upcoming technological innovation. The necessary technological advancements for 6G and beyond are expected to be more intelligent, more efficient, and more secure. The semantic communication paradigm is expected to play a crucial role in the 6G and beyond, as it can provide a more efficient and intelligent way to transmit information. The semantic communication paradigm can be used to reduce the amount of data transmitted, improve the accuracy of the transmitted information, and enable more intelligent communication between devices. In Section 1.3, we discuss the potential applications of semantic communications in the 6G and beyond.

Semantic communications can support a wide range of 6G use cases as following table,[1] for example, multi-sense services, extended reality (XR), human-to-machine communications, and so forth. What these use cases have in common is intensive computation, accurate generation, and reliable communication. Due to the paradigm of semantic communications, the integration between computation, communication, and generation can effectively support these use cases. Take the XR as an example, virtual reality is generated by human interaction and requires cooperation between the edge server and the user's device. In this case, the semantic communications can extract the accurate semantics behind the human language, then transmit the meanings to the edge, and finally receive the generated data, for example, three-dimensional (3D) scenarios, from the edge. Semantic communications can reduce the data transmission and improve the accuracy of generated data. Besides, semantic communications can also support multi-sensory services, for example, multi-sensory holographic teleportation, multi-sensory haptic communications for virtual and augmented reality, and multi-sensory affective computing. These services require the accurate transmission of the human senses, for example, vision, hearing, and touch. Semantic communications can extract the accurate semantics behind the human senses, then transmit the meanings to the edge, and finally receive the generated data, for example, holographic teleportation, haptic communications, and affective computing, from the edge. Semantic communications can reduce data transmission and improve the user experience. In addition, semantic communications can also support human-centric AI, brain-to-computer interactions, and wireless robotics. These services require the accurate transmission of the human's intentions, for example, the commands, the thoughts, and the actions. Semantic communications can extract the accurate semantics behind the human's intentions, then transmit the meanings to the edge, and finally receive the generated data, for example, the AI support, the brain–computer interactions, and the robotics, from the edge. Semantic communications can improve the accuracy of interactions.

Use Case	Service
Multi-sense services	URLLCCC, **Semantic**
Multi-sensory and mobile immersive XR	URLLCCC, **Semantic**
Multi-sensory holographic teleportation	URLLCCC, **Semantic**

1 Machine Type Communications supporting Distributed Intelligence (MMTCCxDI); Geographically Mobile Broadband (GeMBB); Ultra Reliable Low Latency Communications and Computing (URLLCCC).

Use Case	Service
Multi-sensory haptic communications for virtual and augmented reality	URLLCCC, **Semantic**
Multi-sensory affective computing	URLLCCC, **Semantic**
Holographic communication & telepresence	URLLCCC, **Semantic**
Remote XR education	URLLCCC, **Semantic**
Consumption of digital experiences over physical products	URLLCCC, **Semantic**
Human-centric AI support	URLLCCC, **Semantic**
Brain-to-computer interactions	URLLCCC, **Semantic**
Wireless robotics	MMTCCxDI, URLLCCC, **Semantic**
Sustainable connectivity and AI support	MMTCCxDI, GeMBB, URLLCCC, **Semantic**
Society 5.0	MMTCCxDI, GeMBB, URLLCCC, **Semantic**
High precision manufacturing, remote monitoring & control	MMTCCxDI, GeMBB, URLLCCC, **Semantic**
Intelligent disaster prediction	MMTCCxDI, GeMBB, URLLCCC, **Semantic**
Smart digital-twin environments	MMTCCxDI, GeMBB, URLLCCC, **Semantic**
Bidirectional intelligence intertwining (natural and/or artificial)	**Semantic**

References

1 C. E. Shannon and W. Weaver, *The Mathematical Theory of Communication*. Champaign, IL, USA: Univ. Illinois Press, 1949.

2 C. Berrou and A. Glavieux, "Near optimum error correcting coding and decoding: turbo-codes," *IEEE Trans. Commun.*, vol. 44, no. 10, pp. 1261–1271, 1996

3 R. Carnap and Y. Bar-Hillel, "An outline of a theory of semantic information," in *Res. Lab. Electron.*, Massachusetts Inst. Technol., Cambridge, MA, USA, RLE Tech. Rep. 247, 1952.

4 P. Basu, J. Bao, M. Dean, and J. Hendler, "Preserving quality of information by using semantic relationships," *Pervasive Mob. Comput.*, vol. 11, pp. 188–202, 2014.

5 A. Maatouk, M. Assaad, and A. Ephremides, "The age of incorrect information: an enabler of semantics-empowered communication," *IEEE Trans. Wireless Commun.*, vol. 22, no. 4, pp. 2621–2635, 2023.

6 M. Kalfa, M. Gok, A. Atalik, et al., "Towards goal-oriented semantic signal processing: applications and future challenges," *Digit. Signal Process.*, vol. 119, pp. 103–134, 2021.

7 J. Dewey, *Experience and Nature*. Courier Corporation, 1958, vol. 471.

8 L. Wittgenstein, *Philosophical Investigations*. John Wiley & Sons, 2010.

9 M. Jeong, B. Kim, and G. G. Lee, "Semantic-oriented error correction for spoken query processing," in *Proc. IEEE Autom. Speech Recognit. Underst. Workshop (ASRU Workshop)*. IEEE, 2003, pp. 156–161.

10 L. Floridi, "Outline of a theory of strongly semantic information," *Minds Mach.*, vol. 14, no. 2, pp. 197–221, 2004.

11 J. Bao, P. Basu, M. Dean, et al., "Towards a theory of semantic communication," in *Proc. IEEE Netw. Sci. Workshop*, West Point, NY, USA, 2011, pp. 110–117.

12 N. Farsad, M. Rao, and A. Goldsmith, "Deep learning for joint source-channel coding of text," in *Proc. IEEE Int. Conf. Acoust., Speech, Signal Process. (ICASSP)*, Calgary, Canada, 2018, pp. 2326–2330.

13 H. Xie, Z. Qin, G. Y. Li, and B.-H. Juang, "Deep learning enabled semantic communication systems," *IEEE Trans. Signal Process.*, vol. 69, pp. 2663–2675, 2021.

14 Y. LeCun, B. Boser, J. S. Denker, D. Henderson, R. E. Howard, W. Hubbard, and L. D. Jackel, "Backpropagation applied to handwritten zip code recognition," *Neural Comput.*, vol. 1, no. 4, pp. 541–551, 1989.

15 J. J. Hopfield, "Neural networks and physical systems with emergent collective computational abilities," *Proc. Natl. Acad. Sci. U. S. A.*, vol. 79, no. 8, pp. 2554–2558, 1982.

16 S. Hochreiter and J. Schmidhuber, "Long short-term memory," *Neural Comput.*, vol. 9, no. 8, pp. 1735–1780, 1997.

17 K. Cho, B. van Merriënboer, C. Gulcehre, et al., "Learning phrase representations using RNN encoder-decoder for statistical machine translation," in *Proc. Conf. Empir. Methods Nat. Lang. Process. (EMNLP)*, Doha, Qatar, 2014, pp. 1724–1734.

18 M. Schuster and K. Paliwal, "Bidirectional recurrent neural networks," *IEEE Trans. Signal Process.*, vol. 45, no. 11, pp. 2673–2681, 1997.

19 A. Vaswani, N. Shazeer, N. Parmar, et al., "Attention is all you need," in *Proc. Adv. Neural Inf. Proces. Syst. (NeurIPS)*, vol. 30, Long Beach, CA, USA, 2017.

20 M. Dehghani, S. Gouws, O. Vinyals, et al., "Universal transformers," in *Proc. Int. Conf. Learn. Represent. (ICLR)*, New Orleans, LA, USA, 2023.

21 A. Dosovitskiy, L. Beyer, A. Kolesnikov, et al., "An image is worth 16x16 words: transformers for image recognition at scale," in *Proc. Int. Conf. Learn. Represent. (ICLR)*, Vienna, Austria, 2021.

22 J. Devlin, M.-W. Chang, K. Lee, and K. Toutanova, "BERT: pre-training of deep bidirectional transformers for language understanding," in *Proc. Conf. N. Am. Chapter Assoc. Comput. Linguistics: hum. Lang. Technol. (NAACL HLT)*, vol. 1, Minneapolis, MN, USA, 2019, pp. 4171–4186.

23 J. Achiam, S. Adler, S. Agarwal, et al., "GPT-4 technical report," *arXiv preprint arXiv:2303.08774*, 2023.

24 I. Goodfellow, J. Pouget-Abadie, M. Mirza, et al., "Generative adversarial nets," in *Proc. Adv. Neural Inf. Proces. Syst. (NeurIPS)*, vol. 27, Montreal, QC, Canada, 2014.

25 M. Mirza and S. Osindero, "Conditional generative adversarial nets," *arXiv preprint arXiv:1411.1784*, 2014.

26 A. Radford, L. Metz, and S. Chintala, "Unsupervised representation learning with deep convolutional generative adversarial networks," *arXiv preprint arXiv:1511.064 34*, 2015.

27 W. Tong and G. Y. Li, "Nine challenges in artificial intelligence and wireless communications for 6G," *IEEE Wireless Commun.*, vol. 29, no. 4, pp. 140–145, 2022.

28 Y. Shi, Y. Zhou, D. Wen, et al., "Task-oriented communications for 6G: vision, principles, and technologies," *IEEE Wireless Commun.*, vol. 30, no. 3, pp. 78–85, 2023.

29 Z. Qin, X. Tao, J. Lu, et al., "Semantic communications: principles and challenges," *arXiv preprint arXiv:2201.01389*, 2021.

30 H. Ye, G. Y. Li, and B.-H. Juang, "Power of deep learning for channel estimation and signal detection in OFDM systems," *IEEE Wireless Commun. Lett.*, vol. 7, no. 1, pp. 114–117, 2018.

31 Z. Qin, H. Ye, G. Y. Li, and B.-H. F. Juang, "Deep learning in physical layer communications," *IEEE Wireless Commun.*, vol. 26, no. 2, pp. 93–99, 2019.

32 H. Sun, X. Chen, Q. Shi, et al., "Learning to optimize: training deep neural networks for interference management," *IEEE Trans. Signal Process.*, vol. 66, no. 20, pp. 5438–5453, 2018.

33 Z. Ji, Z. Qin, and X. Tao, "Meta federated reinforcement learning for distributed resource allocation," *IEEE Trans. Wireless Commun.*, vol. 23, no. 7, pp. 7865–7876, 2023.

2

Semantic Information Theory

In 1948, the emergence of classical information theory (CIT) was a pivotal milestone, including the fundamental concepts of entropy, channel capacity, and coding theorem. Following the CIT, modern communication techniques have been developed rapidly from first generation (1G) to fifth generation (5G), which are widely applied in various fields. However, CIT is limited to syntactic information, which is insufficient to capture the meaning of information. Semantic information theory (SIT) was proposed to quantify the meaning of information, which is a new research field, to address this limitation. Researchers have made significant progress in SIT since decades ago and have developed various types of SIT based on different mathematical tools. In this chapter, the basic concepts of SIT are introduced, including semantic entropy, semantic channel capacity, and semantic source coding theorem.

2.1 Semantic Entropy

In recent decades, significant efforts have been dedicated to quantifying semantic entropy, inspired by Shannon's foundational work on information entropy. Despite these endeavors, the field remains an active area of research with substantial opportunities for further exploration. This section introduces several distinct definitions of semantic entropy.

2.1.1 Logical Probability Based

Carnap et al. [1] pioneered the development of the classical semantic information theory (CSIT), which relies on logical probabilities to analyze propositional content. Within this framework, sentences are classified as logically true, logically false, or logically indeterminate. For instance, given two sentences i and j,

Deep Learning Enabled Semantic Communications, First Edition. Zhijin Qin et al.

Published 2026 by John Wiley & Sons, Inc.

i entails j if and only if the conditional statement "if i then j" holds as a logical truth. Building upon these logical connections, semantic entropy, a measure of uncertainty in meaning, is formally established, as detailed in Definition 2.1.

Definition 2.1: Let $Pr(x)$ denote the logical probability of a message x, the semantic entropy of x is defined by

$$H_s(x) = -\log_2 (Pr(x)). \tag{2.1}$$

Consider propositions i and j with the following logical probability conditions:

- The truth values of i and j are independently determined by fair coin flips, meaning each has a 50% chance of being true or false.
- The logical probability of the conjunction "$i \wedge j$" being true is 0.25 (occurring in one of four possible scenarios).
- The logical probability of the disjunction "$i \vee j$" being true is 0.75 (occurring in three of four scenarios).

The semantic information content of a statement i is defined as $H_s(i) = -\log_2(Pr(i))$, where $Pr(i)$ is its logical probability. Applying this Definition 2.1, the semantic entropy can be obtained by

- $H_s(a) = H_s(b) = -\log_2(0.5) = 1$
- $H_s(a \wedge b) = -\log_2(0.25) = 2$
- $H_s(a \vee b) = -\log_2(0.75) \approx 0.415$

These results align with the intuitive hierarchy of information, where more specific claims (e.g., conjunctions) carry greater semantic content. Furthermore, logical entailment does not increase semantic information, as shown by the principle $A \models B \Rightarrow H_s(A) \leq H_s(B)$. This formalizes the idea that if A logically implies B, B cannot contain more information than A.

However, Carnap's theory ignored a critical logical contradiction, specifically, the relation "i and not i," which paradoxically yields maximal semantic information rather than minimal. Bao et al. [2] filled this gap by introducing the *world model* to expand CSIT. Their framework introduces the world model, inference mechanism, and message generator to measure the semantic entropy of a given source, which is illustrated as follows.

- The world model is defined as $W_s = \{\boldsymbol{w}_1 \sim p_1, \boldsymbol{w}_2 \sim p_2, \cdots, \boldsymbol{w}_N \sim p_N\}$, in which $\boldsymbol{w}_n$ is interpretation consisting of a set of propositions, and p_n is a logical probability distribution.
- The inference mechanism, I_s, checks the propositional logic satisfiability relation.

- The message generator, M_s, to generate messages based on the world model, for example $M_s(w) = x$, in which w is observed interpretation of a world model and x is the generated message. The relationship $w \vDash x$ must hold, where $\vDash$ represents the conventional propositional logic satisfaction relation, as confirmed by I_s.
- The background knowledge base K, which helps to understand the generated message.

Based on the above assumptions, they introduced the world model-based logical probability of a message (sentence) x, which is defined as

$$p(x|K) = \frac{\sum_{w \in W, w \vDash K, x} p(w)}{\sum_{w \in W, w \vDash K} p(w)}, \tag{2.2}$$

where $W_x = \{w \in W \mid w \vDash x\}$.

With the refined logical probability, the semantic entropy is defined in Definition 2.2.

Definition 2.2: Based on the introduced world model, the semantic entropy of a message x is defined as

$$H_s(x) = -\log_2 p(x|K). \tag{2.3}$$

Compared with the CSIT, Bao's theory introduces the world model, that is, the background described in Definition 2.2. This modeling is closer to human thinking. Therefore, the measurement satisfies some common-sense requirements. Besides, in contrast to CSIT, the refined semantic entropy is based on the conditional logical probability, in which the background knowledge is analogous to the learned knowledge in a human's brain.

2.1.2 Synonymous Mapping Based

Different from the logical probability-based semantic entropy, Niu et al. [3] proposed a synonymous mapping-based semantic entropy. The synonymous mapping is shown in Figure 2.1, in which the synonymous is not only limited to the text but also the image and speech. The synonymous mapping-based semantic entropy is defined as follows.

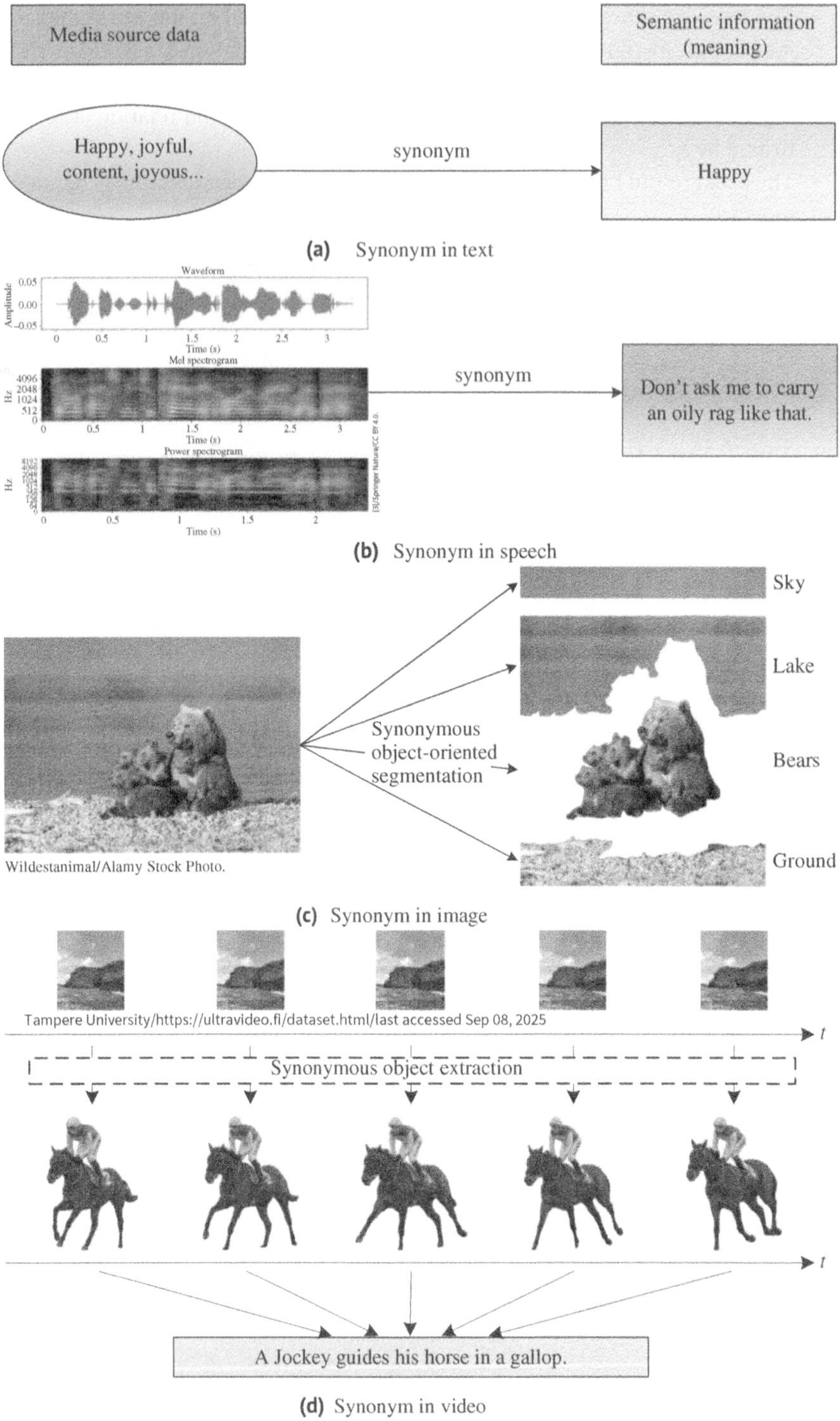

Figure 2.1 Synonymous mapping examples. *Source*: Adapted from [3].

Definition 2.3: Given a continuous random variable X with a probability density distribution $p(x)$, where $x \in \Omega$, and a synonymous mapping $f : \tilde{\mathcal{X}} \to \Omega$, the semantic entropy of $\tilde{X}$ is defined as follows:

$$H_s(\tilde{X}) = \underbrace{-\int_{\Omega} p(x) \log\, p(x)\, dx}_{\text{Differential entropy}} - \underbrace{\mathbb{E}(\log\, L)}_{\substack{\text{Expected log-lengths of} \\ \text{synonymousintervals}}}, \tag{2.4}$$

where $\mathbb{E}(\log L) = \sum_{i_s=-\infty}^{\infty} \int_{\Omega_{i_s}} p(x)\, dx \log L_{i_s}$ represents the expected value of the logarithm of the synonymous interval lengths.

From Definition 2.3, the synonymous mapping-based semantic entropy consists of the differential entropy (i.e., the classical information entropy) and the expected log-lengths of synonymous intervals introduced by the synonymous mapping. Therefore, the synonymous mapping-based semantic entropy is less than or equal to the classical information entropy due to the expected log-lengths of synonymous intervals, which is $H_s(\tilde{X}) \leq H(X)$. This indicates that the synonymous mapping-based semantic source coding can compress the source more efficiently than the classical source coding. Similarly, the semantic conditional entropy satisfies $H_s\left(\tilde{Y}|X\right) \leq H\left(Y|X\right)$. With the above inference, the chain rule of entropies with two variables can be derived as follows.

Theorem 2.1: *(Chain Rule of Entropies with Two Variables):*

$$H_s(\tilde{X}) + H_s(\tilde{Y}\,|X) \leq H_s(\tilde{X}, \tilde{Y}) \leq H(Y) + H_s(\tilde{X}\,|Y) \leq H(X, Y). \tag{2.5}$$

2.1.3 Fuzzy Theory Based

Fuzzy systems [4] introduce the matching degree and the membership degree to characterize the semantic entropy. The membership degree, which ranges between 0 and 1, measures the extent to which an element belongs to a fuzzy set. Values closer to 1 signify stronger membership in a specific class, while lower values indicate weaker affiliation. Specifically, given a semantic concept w, Liu et al. [4] formalized a semantic entropy as

$$H_{C_j}(w) = -D_j(w) \log_2 D_j(w), \tag{2.6}$$

where $D_j(w)$ is the matching degree and is defined as

$$D_j(w) = \frac{\sum_{X \in \mathcal{X}_{C_j}} \mu_w(X)}{\sum_{X \in \mathcal{X}} \mu_w(X)}, \tag{2.7}$$

where $\mu_w(X)$ is the membership degree for each $X \in \mathcal{X}$, and C_j is the class j.

The total semantic entropy across $\mathcal{X}$ is obtained by aggregating the entropies of all classes. This semantic entropy has a similar expression to the classical information entropy. However, it uniquely incorporates membership degrees to capture semantic relevance. Besides, the formulation is similar to the logical probability-based semantic entropy. Both semantic entropies measure the ratio of the given X in the entire model or set. The difference lies in the definition of the entire model. One is the predefined world model, and the other one is the predefined classification set.

2.1.4 Task Based

Given a task, the definition of semantics is the most related information behind the message to the task. Therefore, the framework of semantic entropy should be versatile enough to accommodate diverse tasks. Given the chain $X \to Y \to V$, Chattopadhyay et al. [5] proposed task-based semantic entropy, which quantifies the minimal number of semantic queries Y required about data X, such that it can successfully execute a target task V. Mathematically, this is formulated as

$$\begin{aligned} &H_s(X; V) = \min_{E} \mathbb{E}_X \left[\| \mathrm{Code}_Q^E(X) \| \right], \\ &\text{s.t.} \quad \forall x, y, \quad p(v|\mathrm{Code}_Q^E(x)) = p(v|x), \end{aligned} \tag{2.8}$$

where $\mathrm{Code}_Q^E(X)$ denotes the queries derived from input x via a semantic encoder E. The objective is to minimize the expected number of these queries while making the conditional probability distribution of V approach the $p(v|x)$. This task-based semantic entropy can be applied to various scenarios, such as image classification, speech recognition, and natural language processing.

2.2 Semantic Channel Capacity

Two definitions of semantic channel capacity will be presented in this section: one based on synonymous mapping and the other on logical probability. The semantic channel capacity can give the maximum transmission rate with arbitrary semantic errors.

2.2.1 Logical Probability Based

Given the chain $S \to X \to Y$, in which S is the source, X is the transmitted message, and Y is the received message. The logical probability-based semantic channel capacity is introduced, which is defined as follows.

Theorem 2.2: *The semantic channel capacity of a discrete memoryless channel is defined as*

$$C_s = \sup_{p(X|S)} \left\{ I(X;Y) - H(S|X) + \overline{H_s(Y)} \right\}, \tag{2.9}$$

where $I(X;Y)$ denotes the mutual information between X and Y, $p(X|S)$ represents the conditional probability distribution of generated message X conditioned on S, and $H(S|X)$ indicates the semantic ambiguity introduced by the semantic coding. $\overline{H_s(Y)} = -\sum p(y)H_s(y), y \in \mathcal{Y}$, defines the average semantic entropy of the received message, where $H_s(y)$ is the semantic entropy of y.

The C_s consists of three terms, including the Shannon channel capacity, the semantic ambiguity introduced by the semantic encoder, and the average semantic entropy of the received message. Therefore, the C_s can be higher or lower than the Shannon channel capacity, depending on the design of the semantic encoder and the interpretive proficiency of the receiver. A higher value of $H(S|X)$ reflects increased semantic ambiguity generated by the semantic encoder, while a higher $\overline{H_s(Y)}$ suggests the receiver's enhanced capacity to correctly decode messages. Consider the source sentence: "*He dropped Steven's parcel on the second floor of the apartment, which has 39 floors with 400 sqm on each floor and is called Sam Building due to the designer, Steven Sam.*" The receiver's objective is to determine *where Steven's parcel is dropped*. Two scenarios are examined:

- **Scenario 1**: $\overline{H_s(Y)} > H(S|X)$, the receiver successfully resolves semantic ambiguity introduced by the encoder. For instance, compressing the source sentence to "*the second floor of Sam Building*" retains sufficient clarity despite high $H(S|X)$. The receiver's strong interpretive ability enables accurate inference, allowing C_s to surpass Shannon capacity.
- **Scenario 2**: $\overline{H_s(Y)} < H(S|X)$, the receiver fails to resolve ambiguity. If the source sentence is compressed to "*He dropped Steven's parcel on the building,*" critical details are lost. The receiver cannot pinpoint the car's location, demonstrating insufficient interpretive skill. Here, C_s falls below the Shannon capacity.

2.2.2 Synonymous Mapping Based

Given the transmission chain, $\tilde{\mathcal{X}} \rightarrow \mathcal{X} \rightarrow \mathcal{Y} \rightarrow \tilde{\mathcal{Y}}$, in which $\mathcal{X}$ and $\mathcal{Y}$ denote the input and output syntactic alphabets, respectively, $\tilde{\mathcal{X}}$ and $\tilde{\mathcal{Y}}$ are the corresponding input and output semantic alphabets, respectively. Besides, the discrete channel can be represented as $p(Y|X)$.

Definition 2.4: Given a discrete memoryless channel $\{\tilde{\mathcal{X}}, \mathcal{X}, \mathcal{Y}, \tilde{\mathcal{Y}}, p(Y|X)\}$, the semantic channel capacity is defined as

$$C_s = \max_{f_{xy}} \max_{p(x)} \left[H(X) + H(Y) - H_s(\tilde{X}, \tilde{Y})\right], \tag{2.10}$$

where the maximum is taken over all possible input distributions $p(x)$ and the jointly synonymous mappings $f_{xy} : \tilde{\mathcal{X}} \times \tilde{\mathcal{Y}} \rightarrow \mathcal{X} \times \mathcal{Y}$.

It is important to note that $H_s(\tilde{X}, \tilde{Y})$ represents the joint semantic entropy between $\tilde{X}$ and $\tilde{Y}$. The relationship between the channel capacity and the semantic channel capacity, as defined in Definition 2.4, is articulated in the following corollary.

In order to explicitly compare the synonymous mapping-based semantic channel capacity, the Shannon channel capacity is given first by

$$C = \max_{f_{xy}} \max_{p(x)} [H(X) + H(Y) - H(X, Y)]. \tag{2.11}$$

Compare (2.10) with (2.11), the only difference is the $H(X, Y)$ and $H_s(\tilde{X}, \tilde{Y})$. From the definition of synonymous mapping-based semantic entropy in Section 2.1, it has $H_s(\tilde{X}, \tilde{Y}) \leq H(X, Y)$. Therefore, the semantic channel capacity is bounded above by the channel capacity, that is, $C \leq C_s$. In other words, it shows that semantic communication can achieve a higher transmission rate than conventional communication.

2.3 Semantic Source Coding Theorem

Semantic source coding is a fundamental problem in SIT, which aims to compress the source message while preserving its semantic content. In this section, two different semantic source coding theorems are introduced: one based on logical probability and the other based on synonymous mapping. The semantic source coding theorem provides the lower bound of the coding rate with arbitrary semantic errors.

2.3.1 Logical Probability Based

Given the message set X and the corresponding smallest set of $\bar{X}$, which satisfies

$$p(\bar{x}) = \sum_{x \leftrightarrow \bar{x}} p(x), \tag{2.12}$$

where $x \in X$ is the message and $\bar{x} \in \bar{X}$ is the unique semantically equivalent message.

Then, the lossless semantic source coding theorem, grounded in logical probability, can be formulated as follows.

Theorem 2.3: *For a semantic source with an interface language X, there exists a coding strategy to generate a semantically equivalent interface language X' with Shannon entropy $H(X') \geq H(\bar{X})$, in which $\bar{X}$ is the smallest subset of X.*

For instance, given n propositions, there exist 2^{2^n} combinations. If satisfying $|X| > 2^{2^n}$, where $|X|$ is the size of X, the entropy reduction will be significant. Assuming messages are constructed using connectives $\vee$, $\wedge$, and n proposition names, each message can be modeled as a grammar tree with the depth d, indicating message length of $O(2^{d^2})$. When the depth d meets the inequality $d \geq \sqrt{(1/2 + 2\log n)^2 + 2n + 1}$, the condition $2^{2^n} < |X|$ is satisfied. This results in a maximum permissible message length of approximately $O(2^{2^{n+1}})$ with lossless compression.

2.3.2 Synonymous Mapping Based

Analogous to the channel asymptotic equipartition property (AEP), the semantic AEP yields the following properties for semantically typical sets:

1. If $(\tilde{x}_1, \tilde{x}_2, \ldots, \tilde{x}_n) \in \tilde{A}_\epsilon^{(n)}$, then $H_s(\tilde{X}) - \epsilon \leq -\frac{1}{n}\log p(\tilde{x}_1, \tilde{x}_2, \ldots, \tilde{x}_n) \leq H_s(\tilde{X}) + \epsilon$, equivalently,

$$2^{-n(H_s(\tilde{X})+\epsilon)} \leq p(\tilde{x}_1, \tilde{x}_2, \ldots, \tilde{x}_n) \leq 2^{-n(H_s(\tilde{X})-\epsilon)}. \tag{2.13}$$

2. $\Pr\left\{\tilde{A}_\epsilon^{(n)}\right\} > 1 - \epsilon$ for sufficiently large n.
3. $(1 - \epsilon)\, 2^{n(H_s(\tilde{X})-\epsilon)} \leq \left|\tilde{A}_\epsilon^{(n)}\right| \leq 2^{n(H_s(\tilde{X})+\epsilon)}$ for sufficiently large n.
4. $R_s = H(X) - H_s(\tilde{X})$.

Based on the above conditions, the semantic source coding theorem is given as follows.

Theorem 2.4: *Given the semantic source $\tilde{X}$ and the syntactic source X with the synonymous mapping $f: \tilde{\mathcal{X}} \to \mathcal{X}$, for any code rate $R > H_s(\tilde{X})$, there exists a sequence of $(2^{n(R+R_s)}, n)$ codes such that, as the code length n becomes sufficiently large, the error probability approaches zero, that is, $P_e^{(n)} \to 0$. Conversely, if $R < H_s(\tilde{X})$, then for any $(2^{n(R+R_s)}, n)$ code, the error probability approaches one as n becomes sufficiently large.*

Compare the synonymous mapping-based semantic source coding theorem with the Shannon source coding theorem; it can be found that the bound of coding rate is $H_s(\tilde{X})$ rather than $H(X)$. This is because the semantic source coding theorem is based on the synonymous mapping, which can compress the source more efficiently than the classical source coding.

2.4 Semantic Channel Coding Theorem

Semantic channel coding, a core principle in SIT, establishes the maximum achievable transmission rate with arbitrary semantic errors. This section presents two distinct semantic channel coding theorems: the first grounded in logical probability and the second leveraging a synonymous mapping strategy.

2.4.1 Logical Probability Based

The logical probability-based semantic channel coding theorem is presented as follows.

Theorem 2.5: *Given the semantic channel capacity,*

$$C_s = \sup_{p(X|W)} \left\{ I(X;Y) - H(W|X) + \overline{H_s(Y)} \right\}. \tag{2.14}$$

For any $\epsilon > 0$ and $R < C_s$, there is a block coding strategy such that the maximal probability of semantic error is $< \epsilon$.

Notice that the C_s can be larger or smaller than the Shannon channel capacity. Therefore, the semantic channel coding rate may be lower than that of conventional channel coding, which results in a longer transmission message.

2.4.2 Synonymous Mapping Based

The synonymous mapping-based semantic channel coding theorem is presented as follows.

Theorem 2.6: *Given the semantic channel $\{\tilde{\mathcal{X}}, \mathcal{X}, \mathcal{Y}, \tilde{\mathcal{Y}}, p(Y|X)\}$, for any code rate $R < C_s$, there exists a sequence of $(2^{n(R+R_s)}, n)$ codes, where the synonymous codeword set has a rate $0 \leq R_s \leq H(X,Y) - H_s(\tilde{X}, \tilde{Y})$. As the code length n becomes sufficiently large, the error probability approaches zero,* that is, $P_e^{(n)} \to 0$.

Conversely, if $R > C_s$, then for any $(2^{n(R+R_s)}, n)$ code, the error probability approaches one as n becomes sufficiently large.

For the synonymous mapping-based semantic channel coding theorem, the channel coding rate is higher than the Shannon channel coding rate due to the larger C_S. This means that semantic communication can achieve a higher transmission rate than conventional communication.

2.5 Information Bottleneck

The Information Bottleneck (IB) method is a framework in information theory that aims to find a compressed representation of a random variable that retains the most relevant information about another variable. Next, several aspects of the IB method are discussed.

2.5.1 Classical Information Bottleneck

Given a joint probability distribution $p(x, y)$, the IB optimization problem can be formulated as follows [6]: find a representation T such that the mutual information $I(T; X)$ is minimized, subject to the constraint that $I(T; Y)$ does not exceed a predetermined threshold D. This can be expressed mathematically as

$$\min_{p(t|x)} I(T; X) \quad \text{subject to} \quad I(T; Y) \geq D \tag{2.15}$$

where $I(T; X)$ is the mutual information between input and representation, $p(t|x)$ is conditional probabilities of the presentation given the input.

Alternatively, this can be formulated using a Lagrangian multiplier β as

$$\mathcal{L} = I(T; X) - \beta I(T; Y) \tag{2.16}$$

where β controls the trade-off between compression (minimizing $I(T; X)$) and pre serving relevant information (maximizing $I(T; Y)$).

In the realm of semantic communication, the IB principle can be utilized to develop encoders and decoders that efficiently compress and transmit information while preserving the semantic content pertinent to the communication task. This methodology ensures that the transmitted information is both efficient and meaningful, thereby enhancing the performance of semantic communication systems. For instance, Barbarossa et al. [7] considered a scenario in which transmitted data is affected by channel noise, meaning the receiver obtains a noisy signal $T + n$, with n denoting the noise. To address this, the authors reformulated the IB optimization problem as

$$\min_{\mathcal{E},\mathcal{D}} I(X; T) + \beta \text{MSE}(Y, \hat{Y}), \tag{2.17}$$

where $\mathcal{E}$ and $\mathcal{D}$ correspond to the encoder and decoder, respectively. The term $\text{MSE}(Y, \hat{Y})$ quantifies the reconstruction error between the true output Y and its estimate $\hat{Y}$, derived as $T = \mathcal{E}(X)$ (encoded representation) and $\hat{Y} = \mathcal{D}(T + n)$ (decoded output). The parameter β balances the trade-off between minimizing the mutual information $I(X; T)$ (complexity) and reducing the reconstruction error (information relevance).

2.5.2 Knowledge Collision-based Information Bottleneck

With the concept of semantic expansion and knowledge collision to the SIT, Xin et al. [8] proposed a novel IB method, which is expressed mathematically as

Lemma 2.1: *Consider a semantic transmission chain $T_A \rightarrow T_B \rightarrow T_C$. If one channel is noiseless and the other channel exhibits conditional Markov properties, then a generalized data-processing inequality at the semantic level can be derived, expressed as*

$$I(T_A; T_C) \leq I(T_A; T_B). \tag{2.18}$$

Lemma 2.1 implies that no processing of T_B at the semantic level, whether deterministic or stochastic, can enhance the information that T_B retains about T_A. Connecting to the IB principle, the knowledge collision-based IB method aims to find a representation T_B that retains the most relevant information about T_A while minimizing the mutual information between T_B and T_C.

References

1 R. Carnap and Y. Bar-Hillel, "An outline of a theory of semantic information," in *Res. Lab. Electron.*, Massachusetts Inst. Technol., Cambridge, MA, USA, RLE Tech. Rep. 247, 1952.

2 J. Bao, P. Basu, M. Dean, et al., "Towards a theory of semantic communication," in *Proc. IEEE Netw. Sci. Workshop*, West Point, NY, USA, 2011, pp. 110–117.

3 K. Niu and P. Zhang, "A mathematical theory of semantic communication," *J. Commun.*, 2024.

4 X. Liu, W. Jia, W. Liu, and W. Pedrycz, "AFSSE: an interpretable classifier with axiomatic fuzzy set and semantic entropy," *IEEE Trans. Fuzzy Syst.*, vol. 28, no. 11, pp. 2825–2840, 2020.

5 A. Chattopadhyay, B. D. Haeffele, D. Geman, and R. Vidal, "Quantifying task complexity through generalized information measures," in *Proc. Int. Conf. Learn. Represent. (ICLR)*, 2011.

6 N. Tishby and N. Zaslavsky, "Deep learning and the information bottleneck principle," in *IEEE Inf. Theory Workshop*, IEEE, 2015, pp. 1–5.

7 S. Barbarossa, D. Comminiello, E. Grassucci, et al., "Semantic communications based on adaptive generative models and information bottleneck," *IEEE Commu. Mag.*, vol. 61, no. 11, pp. 36–41, 2023.
8 G. Xin, Z. Zhu, and P. Fan, "Information framework expansion meets knowledge collision for semantic communications," in *IEEE Int. Conf. Commun.*, IEEE, 2023, pp. 6096–6101.

3

Joint Semantic-channel Coding for Source Reconstruction

This chapter introduces deep learning (DL)-enabled semantic communications for different source reconstruction tasks. In particular, the structure of joint semantic-channel coding for source reconstruction is shown in Figure 3.1. From the figure, various neural network (NN)-based semantic encoders are tailored for different sources to extract semantic features. These semantic features are mapped into transmittable symbols through the channel encoder and then transmitted over the wireless channel. At the receiver, the received symbols are decoded by the channel decoder to recover the semantic features, which are then fed into a semantic decoder to reconstruct the original source. Generally, the semantic encoder and decoder, as well as the channel encoder and decoder, are designed in a symmetric structure. The final reconstructed source aims to be as identical as possible to the original source. More importantly, the semantic encoding and channel encoding are jointly devised and optimized, enabling global end-to-end (E2E) performance while avoiding unnecessary computational overhead.

According to diverse sources, there are three individual sections to discuss the use cases of text, speech, and image/video reconstruction. Each section includes details on signal processing techniques for specific sources and the system model of the considered DL-enabled semantic communications. Additionally, the performance metrics for measuring the quality of the reconstructed sources and related experimental results are provided to demonstrate the superiority of semantic communications for generating the original data compared to current conventional communication systems.

Deep Learning Enabled Semantic Communications, First Edition. Zhijin Qin et al.

Published 2026 by John Wiley & Sons, Inc.

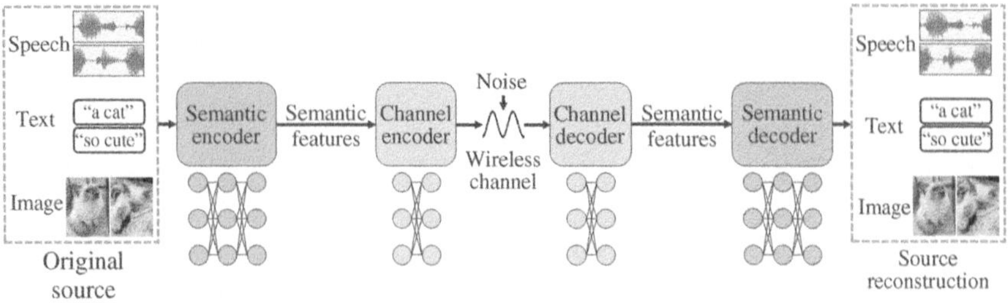

Figure 3.1 The structure of joint semantic-channel coding for source reconstruction.

3.1 Semantic Communications for Text

Semantic communication considers the difference between the meaning of the transmitted messages and that of recovered ones, for example, the semantic errors. However, the difference cannot be captured by the conventional separate coding methods due to the lack of ability to understand text. Natural language processing (NLP) paved the path to understanding the text by employing the NN to learn the latent semantic features.

NLP is a computer program's ability to understand human language. It's a part of artificial intelligence (AI) and uses techniques from computer science, linguistics, and AI. NLP can analyze and generate human language, including speech and text. It can also translate text from one language to another. Basic knowledge about DL-based NLP, which includes tokenizers, embedding, and NNs, will be introduced. Given the example sentence, "We are learning NLP today," this knowledge will be introduced one by one.

The first is **tokenizer**, which splits the input sentence into the tokens according to the predefined codebook with size l. The token could be a word or sub-word. For example, the sentence can be split into two cases,

- Word tokens: ["we," "are," "learning," "NLP," "today"].
- Sub-word tokens: ["we," "are," "learn," "ing," "NLP," "today"].

After tokenization, the word or sub-word will be transformed into the learnable vector by **embedding** technique, such that the computer can process the words. First, the learnable matrix $\boldsymbol{E} \in \mathbb{R}^{l \times d}$ can be built, each vector of which refers to the unique word or sub-word in the predefined codebook. Then, the word or sub-word tokens can be mapped to the vectors of the embedding matrix. For example, the word tokens ["we," "are," "learning," "NLP," "today"] can be mapped to $\boldsymbol{E}_1 \in \mathbb{R}^{5 \times d}$.

Next, NNs are designed to process vectors, in which the structure can be categorized as the **recurrent neural network (RNN)** and **transformer**. The comparisons are given in the following table.

Feature	RNN	Transformer
Processing Method	Sequential Processing	Parallel Processing
Long-range Dependency	Prone to Attenuation	Directly Modeled by Self-attention
Computational Complexity	O(n)	O(n^2) (Attention Matrix)
Positional Information	Implicitly Included	Explicit Positional Encoding

3.1.1 Joint Semantic-channel Coding for Text

Given the input sequence, $\boldsymbol{s} = [w_1, w_2, \ldots, w_N]$, where w_n corresponds to the n-th word in the sequence. At the transmitter, two components are employed: a semantic encoder and a channel encoder. These work together to extract semantic features from $\boldsymbol{s}$ and ensure robust transmission over physical channels. The transmitted symbol stream is generated as

$$\boldsymbol{x} = C_{\alpha}\left(S_{\beta}(\boldsymbol{s})\right), \tag{3.1}$$

where $S_{\beta}(\cdot)$ and $C_{\alpha}(\cdot)$ represent semantic encoder and channel encoder, respectively. $\boldsymbol{\alpha}$ and $\boldsymbol{\beta}$ are the learnable parameters.

After transmission, the received signal at the receiver is modeled as

$$\boldsymbol{y} = h\boldsymbol{x} + \boldsymbol{n}, \tag{3.2}$$

where h is the channel gain and $\boldsymbol{n} \sim \mathcal{CN}(0, \sigma_n^2)$ is additive complex Gaussian noise.

In order to recover the sentence with the same meaning, the receiver employs a channel decoder and a semantic decoder to reconstruct the original sentence, which is expressed as

$$\hat{\boldsymbol{s}} = S_{\chi}^{-1}\left(C_{\delta}^{-1}(\boldsymbol{y})\right), \tag{3.3}$$

where $\hat{\boldsymbol{s}}$ is the reconstructed sentence, $C_{\delta}^{-1}(\cdot)$ and $S_{\chi}^{-1}(\cdot)$ are the channel encoder and decoder, respectively. χ and δ are the learnable parameters.

Figure 3.2 presents the structure of the deep joint semantic channel coding, named DeepSC. The structure takes the deep neural networks (DNNs) to design both semantic and channel codecs, which is beneficial in E2E training. Specifically, the physical channels are interpreted as one untrained layer in the model. The transmitter takes the $\boldsymbol{s}$ as the input. With the NLP techniques, $\boldsymbol{s}$ is first mapped to word embeddings by word-to-vector methods and input to multiple transformer encoder layers to extract the semantic features behind the texts. Then, the channel encoder consists of multiple dense layers with different widths, which work to generate the $\boldsymbol{x}$ and protect it over the air. Accordingly, the receiver is composited

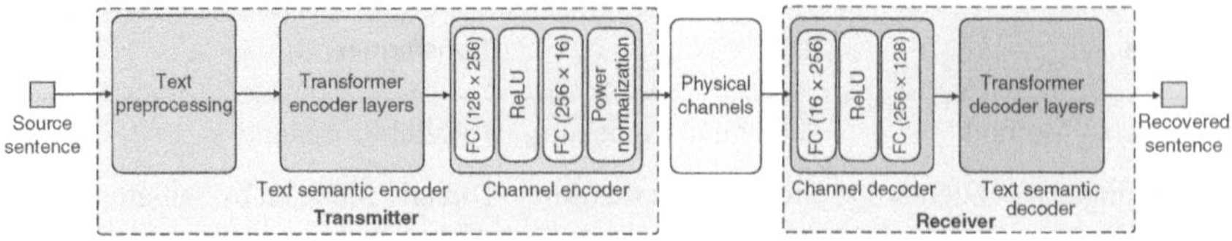

Figure 3.2 The structure of the DeepSC. *Source*: Adapted from [1].

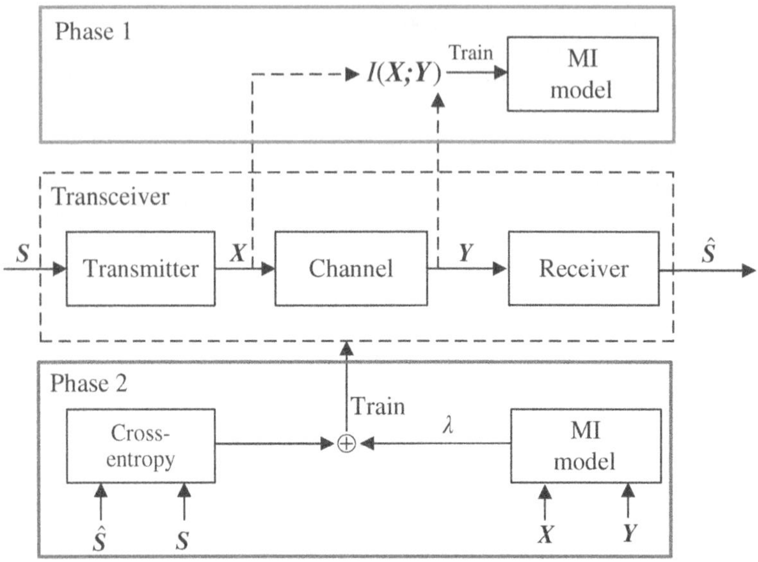

Figure 3.3 Two-stage training for the DeepSC. *Source*: [1]/IEEE/CC BY 4.0.

with the dense layers-based channel decoder to estimate semantic features from the received signals and the transformer decoder layers-based semantic decoder to recover $\boldsymbol{s}$ from the semantics. Such a design is beneficial for reducing complexity and solving the difficulty of training the network because semantic and channel codecs adopt different substructures instead of directly employing the same DNNs to build it.

The implementation of the DeepSC is shown in Figure 3.3, including two stages, that is, mutual information (MI) estimation model training and DeepSC training stages. Given the training dataset, during the MI estimation model training, $\boldsymbol{x}$ and $\boldsymbol{y}$ are fed into the MI model [2] to update its parameters, which learn to provide an accurate MI estimation. The MI can provide extra information to train a receiver, maximizing the capacity or the data transmission rate. During the DeepSC training stage, both the MI estimated by the MI model and the semantic difference between

$\boldsymbol{s}$ and $\hat{\boldsymbol{s}}$ are employed to train DeepSC. Hence, the model learns to improve the semantic errors and transmission rate together. Besides, due to the parameters of DeepSC updates after training, the latent distributions of $\boldsymbol{x}$ and $\boldsymbol{y}$ are also changed. The MI model cannot estimate the accurate MI value based on the parameters obtained by the past $\boldsymbol{x}$ and $\boldsymbol{y}$. It is critical to update the MI model to correct the estimation errors. Therefore, the iterative training is employed to implement the DeepSC. Notably, the channel realization is randomly sampled from the Additive White Gaussian Noise (AWGN) channel with certain statistical properties, which functions to generate random samples for training. In terms of the loss function during training, cross-entropy (CE) and MI can be used to evaluate the network partially. In [1], the model is trained by minimizing CE and maximizing MI

$$\mathcal{L}_{\text{total}} = \mathcal{L}_{\text{CE}}(\boldsymbol{s}, \hat{\boldsymbol{s}}) - \lambda \mathcal{L}_{\text{MI}}(\boldsymbol{x},\boldsymbol{y}), \tag{3.4}$$

where $\mathcal{L}_{\text{CE}}$ and $\mathcal{L}_{\text{MI}}$ are CE and MI loss, respectively. λ is the coefficient to balance the $\mathcal{L}_{\text{CE}}$ and $\mathcal{L}_{\text{MI}}$. After sufficient training, the coefficients and weights in the model are fixed, and the DeepSC is deployed online for forward inference.

Figure 3.4 illustrates the word error rate (WER) performance comparison between DeepSC and conventional communication systems over AWGN channels. Conventional approaches employ Huffman coding or fixed-length coding for source encoding, paired with Turbo coding for channel encoding. In contrast,

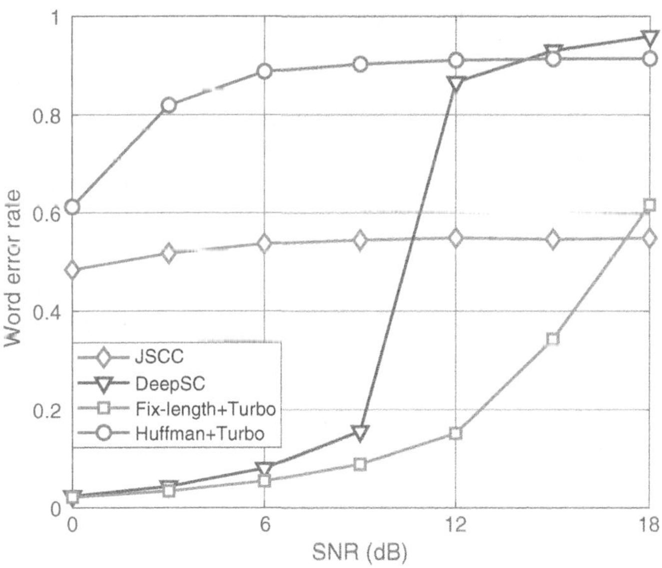

Figure 3.4 WER as SNR changes for the same number of symbols transmitted over the AWGN channels.

JSCC integrates LSTM networks and dense layers to map textual data directly to transmitted bit streams, jointly optimizing both coding processes. The results highlight two key observations: First, DeepSC significantly outperforms all benchmarks in low SNR scenarios and nearly matches the WER of Huffman–Turbo coding at high SNRs, demonstrating its enhanced robustness in noisy environments. Second, Huffman–Turbo coding exhibits a marginally higher WER than DeepSC. This discrepancy stems from conventional methods' property of precise transmission, which limits their adaptability under varying channel conditions compared to the flexible, learning-driven approach of DeepSC.

However, the WER only measures the difference between texts and ignores the semantic differences behind the text. For example, the two sentences "I have a car" and "I have an automobile" have the same meaning but with a high WER. Therefore, semantic-related metrics are required. Section 3.2.1 will introduce several performance metrics to measure semantic errors in text.

3.2 Semantic Communications for Speech

This section introduces a DL-enabled semantic communication system for speech transmission, DeepSC [3]. Speech is a significant source in communication networks and shows uniqueness and higher complexity compared to text, which contains not only textual information but also acoustic information. Speech signals consist of a sequence of speech samples that can be a positive number, a negative number, or zero. In conventional communications, speech signals are typically captured by external microphone devices and converted to digital signals through sampling, quantization, and encoding. In addition, conventional speech coding algorithms aim to represent speech samples with the minimum number of binary bits by exploring the dependencies and statistical properties of these samples. Such bit-oriented coding mechanisms ignore the semantics inherent in the speech signals and increase their vulnerability in harsh transmission environments. For example, the speech signals received at chaotic airports and train stations are usually corrupted. To tackle this obstacle, the DL techniques provide a promising solution to explore semantic information and achieve semantic exchange for speech signals. This section presents Convolutional Neural Network (CNN)-based DeepSC-S proposed in [3]. Compared with conventional speech transmission systems, DeepSC-S achieves more accurate speech reconstruction and shows higher robustness to channel variations.

Denote speech signals as $\boldsymbol{S} = [S_1, S_2, ..., S_W]$ with W speech samples, where S_w is the w-th element in $\boldsymbol{S}$. The amplitude of speech signals can be affected by various factors, such as the loudness of speakers, the noise level of environments, and the sensitivity of microphone devices. To ensure the consistency of subsequent processing, speech signals are subjected to normalization to the constraint that all speech samples fall within the same range and eliminate the effects of amplitude

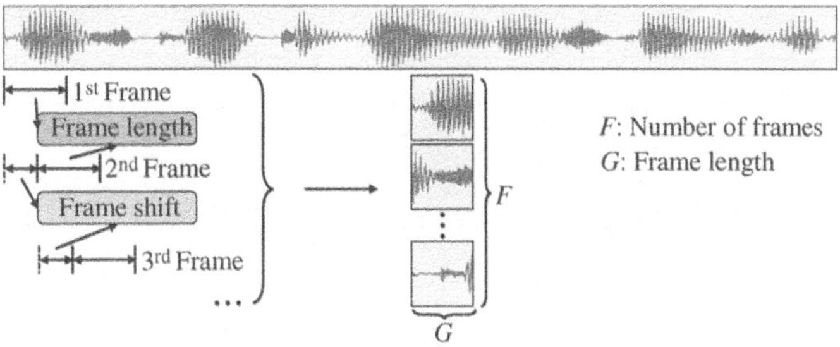

Figure 3.5 An example of the speech framing operation.

discrepancies. In general, the Euclidean normalization is applied to speech signals, expressed as

$$\boldsymbol{S}_{\text{norm}} = \frac{\boldsymbol{S}}{\sqrt{S_1^2 + S_2^2 + \cdots + S_w^2}}. \tag{3.5}$$

Furthermore, speech signals are variable in length due to differences in spoken content, speaker speed, and natural pauses, which impose challenges in data alignment and batch processing. To this end, variable-length speech sequences are divided into multiple frames, as shown in Figure 3.5. The framed speech is denoted as $\boldsymbol{M} \in \mathbb{R}^{B \times F \times G}$, where F is the number of the frames and G represents the length of each frame. It is noteworthy that the framing operation does not change the amplitude of the speech samples. In practice, the frame length is typically selected within the range of 20–40 ms, while the frame stride is commonly set between 10 and 20 ms.

3.2.1 Joint Semantic-channel Coding for Speech

Conventional source-channel coding architecture maps the input speech into the bit sequence prior to wireless transmission. However, in DeepSC-S, the speech is first condensed to remove the redundancy of semantic-agnostic information and converted to floating-point value-based semantic features by sophisticated NNs. These semantic features are encoded into modulated symbols using NN-enabled channel coding. A generic model of the joint semantic-channel coding for speech transmission is shown in Figure 3.6. From the figure, the semantic features, $\boldsymbol{F}$, are obtained through the NN-based semantic encoder.

It is obvious that during silent periods, speech signals carry no meaningful information, whereas speech segments during speaking periods contain significant content. Consequently, speech samples with zero magnitudes should be disregarded, while those with high magnitudes require more attention. Additionally, speakers tend to raise their voices to emphasize important information, leading to a sharp increase in the magnitude of the speech signals during these periods.

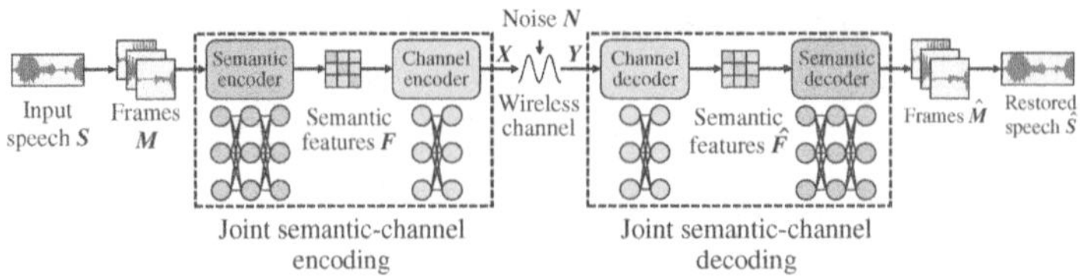

Figure 3.6 The model structure of the DL-enabled semantic communication system for speech reconstruction.

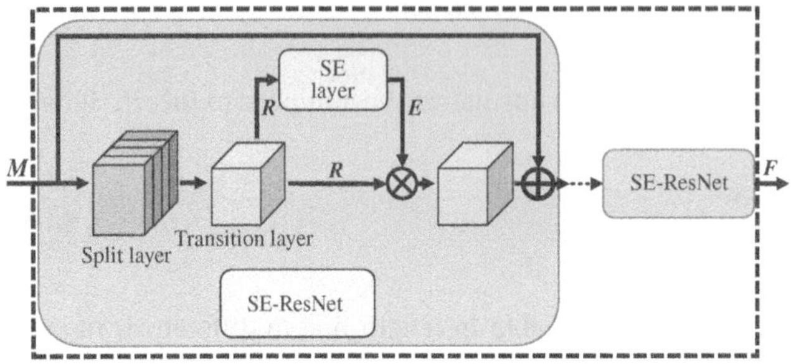

Figure 3.7 Attention-based semantic encoder.

On the other hand, when speaking slowly to convey difficult-to-understand information, the frequency of speech signals tends to decrease abruptly. The essential information refers to the features corresponding to the speaking periods. Inspired by this, an attention mechanism based on squeeze-and-excitation (SE) networks [4], termed SE-ResNet, is utilized to design the semantic encoder, as shown in Figure 3.7. From the figure, the SE layer returns a coefficient vector, $\boldsymbol{E}$, to scale the importance of different semantic features and control the high coefficients assigned to the essential features. The scaled semantic features are fed into the CNN-based channel encoder to produce the symbols, $\boldsymbol{X}$. Denote the semantic encoder and channel encoder as $\mathfrak{T}_S(\cdot)$ and $\mathfrak{T}_C(\cdot)$, respectively, then the encoded symbols are expressed as

$$\boldsymbol{X} = \mathfrak{T}_C(\mathfrak{T}_S(\boldsymbol{M})) \quad \text{w.r.t.} \quad \boldsymbol{\alpha}, \tag{3.6}$$

where $\boldsymbol{\alpha}$ is the joint NN parameters of $\mathfrak{T}_S(\cdot)$ and $\mathfrak{T}_C(\cdot)$.

The symbols, $\boldsymbol{X}$, are normalized to constrain the transmission power such that $\mathbb{E}\|\boldsymbol{X}\|^2 = 1$. And the received symbols, $\boldsymbol{Y}$, can be modeled as

$$\boldsymbol{Y} = \boldsymbol{H}\boldsymbol{X} + \boldsymbol{N}, \tag{3.7}$$

where $\boldsymbol{H}$ denotes the fading channel and $\boldsymbol{N}$ is the AWGN.

The joint semantic-channel decoding commits to recovering accurate semantic features and speech frames, $\widehat{\boldsymbol{M}}$. The channel decoder consists of multiple CNNs, and the semantic decoder is constructed by cascaded SE-ResNet modules. Denote the channel decoder and semantic decoder as $\mathfrak{T}_{\mathrm{C}}^{-1}(\cdot)$ and $\mathfrak{T}_{\mathrm{S}}^{-1}(\cdot)$, respectively. The recovered $\hat{\boldsymbol{M}}$ is denoted as

$$\hat{\boldsymbol{M}} = \mathfrak{T}_{\mathrm{S}}^{-1}(\mathfrak{T}_{\mathrm{C}}^{-1}(\boldsymbol{Y})) \quad \text{w.r.t.} \quad \boldsymbol{\beta}, \tag{3.8}$$

where $\boldsymbol{\beta}$ indicates the trainable parameters. The frames, $\hat{\boldsymbol{M}}$, are concatenated to obtain the speech sequence $\hat{\boldsymbol{S}}$.

The whole speech transmission process is considered as the waveform reconstruction task to minimize the error between $\boldsymbol{S}$ and $\hat{\boldsymbol{S}}$. Moreover, to enable E2E training for the whole semantic communication system, the accurate channel state information (CSI) is assumed to backpropagate the training gradients from the receiver to the transmitter. The mean-squared error (MSE) can be adopted as the loss function to update NN parameters, expressed as

$$\mathcal{L}_{\mathrm{MSE}}(\boldsymbol{S}, \hat{\boldsymbol{S}}; \boldsymbol{\alpha}, \boldsymbol{\beta}) = \sum_{w=1}^{W} (S_w - \hat{S}_w)^2. \tag{3.9}$$

Two benchmarks are recommended for performance comparison to demonstrate the superiority of the DeepSC-S. The first benchmark is the conventional system that transmits the input speech by adopting the pulse code modulation (PCM) codes [5], the Turbo codes [6], and the 64-QAM as the source coding, channel coding, and modulation, respectively, as shown in Figure 3.8.

The second benchmark is named the semi-conventional system, which utilizes the feature encoder to generate the features and uses the conventional system to transmit and recover these features, as shown in Figure 3.9. The feature codecs include multiple convolutional layers to extract the floating-point value-based feature vectors.

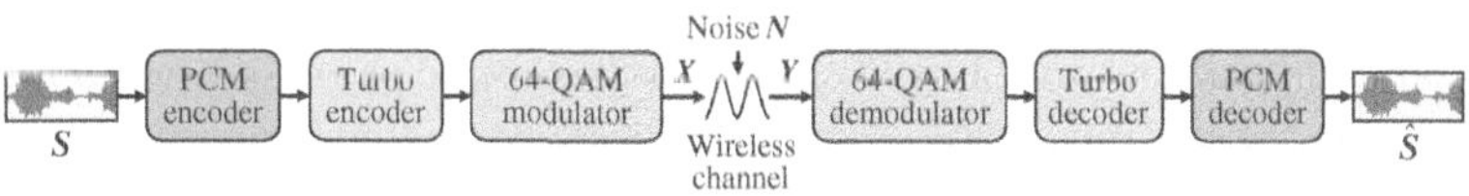

Figure 3.8 The model structure of the conventional system.

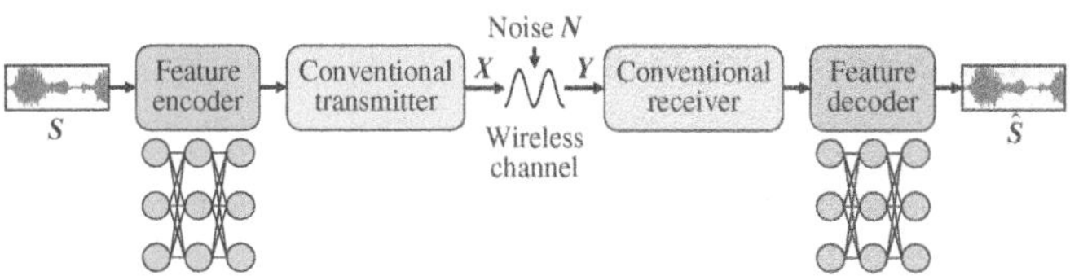

Figure 3.9 The model structure of the semi-conventional system.

The comparison of the perceptual evaluation of speech quality (PESQ) [7] scores is illustrated in Figure 3.10, where the CNN system refers to replacing the SE-ResNet modules in DeepSC-S with CNN modules. From the figure, both the CNN system and DeepSC-S outperform the other two benchmarks across different fading channels and SNRs and demonstrate the ability to mitigate the cliff effect. Moreover, DeepSC-S exhibits robust PESQ performance in contending with channel variations, whereas conventional approaches show poor results at low SNR regimes. DeepSC-S also achieves higher scores than the CNN system, with an average improvement of 7.34% in PESQ, which proves the effectiveness of the attention mechanism. These results indicate that DeepSC-S enhances speech transmission quality under complex communication scenarios more effectively than conventional methods by extracting and transmitting essential semantic information.

3.3 Semantic Communications for Image

Images are more complex due to their high dimensionality and spatial dependencies compared to text and speech. Images contain considerable redundant information, such as pixel-level details. This leads to difficulties when designing the semantic coding mechanism, including extracting key content for compression, maintaining semantic consistency, and learning temporal information.

Image processing uses the computer to process the image data, in which the input is an image and the output is an image or a set of characteristics or parameters related to the image. DL revolutionized image processing. It made previous techniques, based on manual feature extraction, obsolete. Given an image, the NN can capture the dependence between pixels, extract the features from the image, and then use these features to perform the desired tasks. In detail, the features can be divided into low-, mid-, and high-level features. The low-level features are the basic features, such as edges, corners, and textures. The mid-level features are the combination of low-level features, such as shapes and patterns. The high-level features are the semantic features, such as objects, scenes, and concepts. The high-level features are the most important features for image processing tasks. By increasing the depth of the NN, the low-level features are transformed into high-level features that are used to perform the desired tasks.

3.3.1 Joint Semantic-channel Coding for Image

3.3.1.1 System Model

Given an image, $\boldsymbol{I} \in \mathbb{R}^{3\times H\times W}$, where H and W are the height and width of the image. The semantic information can be extracted by

$$\boldsymbol{z} = S(\boldsymbol{I}; \boldsymbol{\alpha}_t), \tag{3.10}$$

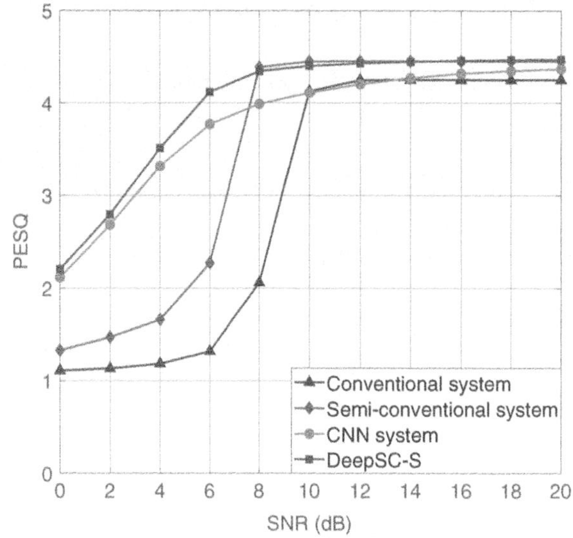

(a) AWGN channels

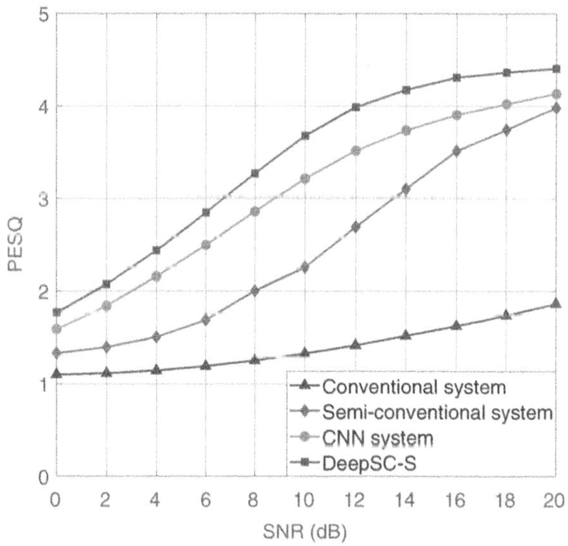

(b) Rayleigh channels

Figure 3.10 PESQ score versus SNR for speech transmission with the conventional system, the semi-conventional system, the CNN system, and the DeepSC-S under AWGN channels and Rayleigh channels: (a) AWGN channels and (b) Rayleigh channels.

where $\boldsymbol{z} \in \mathbb{R}^{M\times 1}$ is the semantic information and $S(\cdot;\boldsymbol{\alpha}_t)$ is denoted as the semantic encoder with the parameter $\boldsymbol{\alpha}_t$.

The encoded symbols for analog semantic transmission are represented as

$$\boldsymbol{x} = C(\boldsymbol{z};\boldsymbol{\beta}_t), \tag{3.11}$$

where $\boldsymbol{x} \in \mathbb{C}^{L\times 1}$ is the encoded complex symbols and $C(\cdot;\boldsymbol{\beta}_t)$ is denoted as the analog channel encoder with the parameter $\boldsymbol{\beta}_t$.

When $\boldsymbol{x}$ is transmitted over the block fading channels, the received signal can be given by

$$\boldsymbol{y} = h\boldsymbol{x} + \boldsymbol{n}, \tag{3.12}$$

where h is the channel coefficient that remains constant within a channel coherence time, $\boldsymbol{n}$ is the noise and follows $\mathcal{CN}\left(0, \sigma_n^2 \mathbf{I}_L\right)$. The SNR is defined as $\mathbb{E}(\|\boldsymbol{x}\|^2)/\mathbb{E}(\|\boldsymbol{n}\|^2)$.

The receiver comprises signal detection that estimates the transmitted symbols, an analog-digital fusion module that fuses the digital and analog semantic information, channel decoders that alleviate the distortions from the wireless channels, and a semantic decoder that recovers the images with the received semantic information.

With channel equalization, the transmitted symbols can be recovered by

$$\hat{\boldsymbol{x}} = \frac{h^*}{|h|^2}\boldsymbol{y} = \boldsymbol{x} + \frac{h^*}{|h|^2}\boldsymbol{n}, \tag{3.13}$$

where $\hat{\boldsymbol{x}}$ represents the estimated symbols. Assume that h is the perfect CSI. After the signal detection, the semantic features are recovered by the analog and digital receivers, respectively.

The semantic features transmitted by analog communications are estimated by

$$\hat{\boldsymbol{z}} = C^{-1}(\hat{\boldsymbol{x}};\boldsymbol{\beta}_r), \tag{3.14}$$

where $\hat{\boldsymbol{z}}$ is the estimated semantic features and $C^{-1}(\cdot;\boldsymbol{\beta}_r)$ is denoted as the analog channel decoder.

Finally, the transmitted image can be reconstructed by

$$\hat{\boldsymbol{I}} = S^{-1}(\hat{\boldsymbol{z}};\boldsymbol{\alpha}_r), \tag{3.15}$$

where $S^{-1}(\cdot;\boldsymbol{\alpha}_r)$ represents the semantic decoder.

3.3.1.2 System Design

An example of joint semantic-channel coding for image, HDA-DeepSC [8], is shown in Figure 3.11. The design of each module is detailed below.

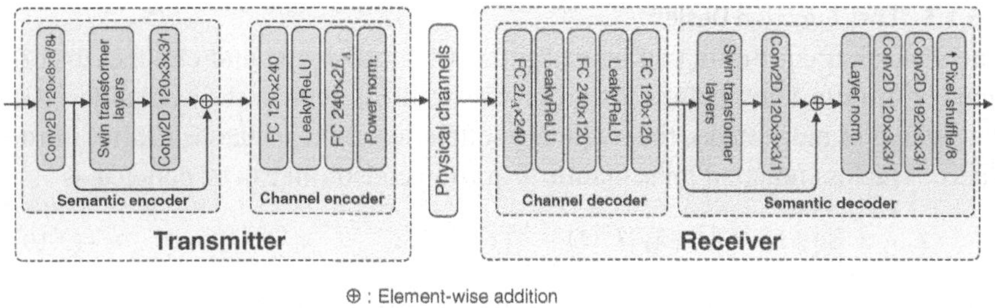

Figure 3.11 The structure of the semantic communication systems for image transmission.

3.3.1.2.1 Semantic Codec

The **semantic encoder** comprises a convolutional layer and a residual Swin transformer block. The first convolutional layer projects the images into vector-shaped tokens, which are used as inputs to the residual Swin transformer block in a permutation-invariant manner. Then, the residual Swin transformer block consists of several Swin transformer layers and a convolution layer, in which the Swin transformer layer originates from the transformer and introduces the local attention and shifted window mechanism to improve the visual semantic understanding. Besides, a convolutional layer with spatially invariant filters in the residual block can enhance the translational equivariance. The residual connection allows for aggregation of the shallow and deep semantic features.

Similarly, the **semantic decoder** consists of the residual Swin transformer block, convolutional layers, and pixel shuffle. The residual block enhances visual semantic understanding. The residual connection provides a short connection from the semantic encoder to the semantic decoder, allowing the processing of reconstruction to fuse varying levels of features. The convolutional layers and pixel shuffle form the reconstruction module, in which the sub-pixel convolutional layer upsamples the feature and pixel shuffle reallocates the features to reconstruct the transmitted images.

3.3.1.2.2 Channel Codec

The channel codec is designed to efficiently compress and transmit semantic features over the air. It primarily utilizes fully connected layers to preserve global semantic information. Unlike convolutional layers, which are adept at capturing local information, fully connected layers excel in maintaining global attributes, aligning with the objectives of the channel codec. This approach enhances the system's robustness to channel noise.

3.3.1.3 Loss Function Design

The distortion constraint can be categorized into semantic and channel distortion constraints. For semantic distortion constraint, except for the pixel difference considered in most works, the frequency difference of the images is further introduced. The loss function for semantic distortion constraint can be denoted as

$$\mathcal{L}_{\mathrm{SD}} = \mathbb{E}\left[\| \boldsymbol{I} - \hat{\boldsymbol{I}} \|^2 + \lambda_{\mathcal{F}} |\mathcal{F}(\boldsymbol{I}) - \mathcal{F}(\hat{\boldsymbol{I}})|\right], \tag{3.16}$$

where $\lambda_{\mathcal{F}}$ is the weight and $\mathcal{F}(\cdot)$ represents the Fourier transform.

The first item in (3.16) measures pixel-wise differences. Assuming image pixels follow a Gaussian distribution, the mean squared error (MSE) loss is applied here. The second item in (3.16) evaluates discrepancies in the frequency domain. By mapping images to their Fourier representations, this term encourages NNs to learn long-range dependencies in the image structure. The reasons behind the design can be summarized as

- The MSE loss guides the NNs to recover the local pixels of the images by comparing the pixel difference, which overlooks global relationships between distant regions in an image.
- The Fourier-based loss can help the NN to learn the long-range dependencies of the image. Because the same frequency in the frequency domain refers to the different pixels at the different positions of the image.

For the channel distortion constraint, the distortions from channels and the transmission of essential information are considered. The designed loss function can be expressed as

$$\mathcal{L}_{\mathrm{CD}} = \mathbb{E}\left[\| \boldsymbol{z} - \hat{\boldsymbol{z}} \|\right], \tag{3.17}$$

where the item minimizes the distortions from channels.

3.3.1.4 Numerical Results

Figure 3.12 reports the peak signal-to-noise ratio (PSNR) and multi-scale structural similarity index (MS-SSIM) comparison between the various methods over AWGN channels with a 1/6 bandwidth compression ratio. It can be observed from Figure 3.12 that the HDA-DeepSC outperforms all the benchmarks. This indicates that the discrete signals of the digital component can accurately deliver crucial semantic information for details recovery, and the continuous signals of the analog component can prevent the leveling off and cliff-edge effects for lower quantization errors. Besides, the HDA-DeepSC achieves the best performance in terms of MS-SSIM, which means that the images transmitted by HDA-DeepSC have better global quality. This is likely because the Fourier-based loss function is introduced,

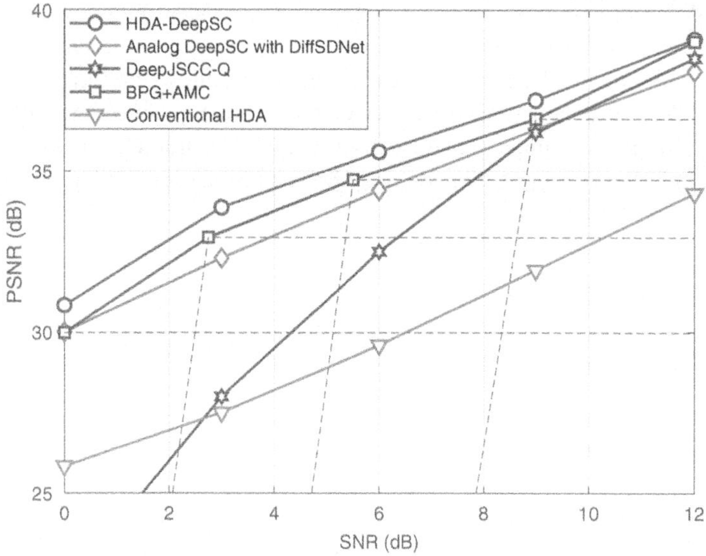

(a) PSNR

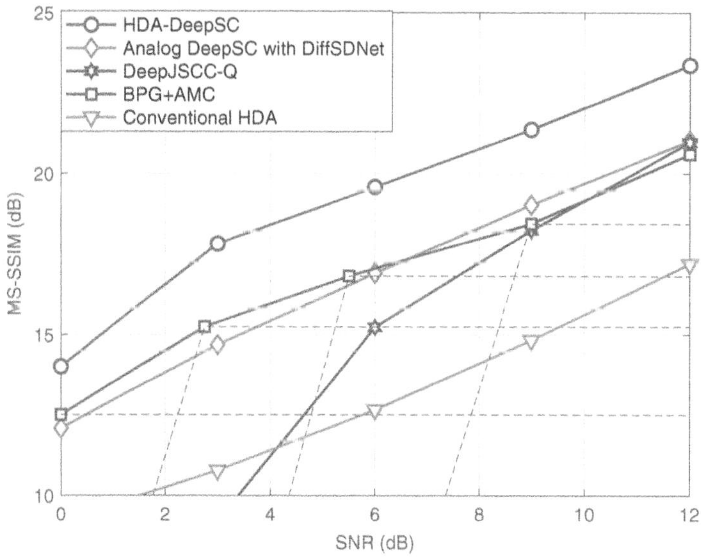

(b) MS-SSIM

Figure 3.12 (a) PSNR and (b) MS-SSIM performance for different methods on the Kodak dataset over AWGN channels.

forcing the model to learn the long-distance dependencies. Besides, if the communication environment is terrible, in which the digital signals cannot be successfully decoded, this system will experience the cliff-edge effect due to the employed entropy coding. This can be improved in several ways. One is to replace the entropy coding module with the learning-based quantization module. Another is to introduce error transmission during training. Both methods can lead the model to learn to correct the errors in digital transmission.

3.4 Performance Metrics

3.4.1 Performance Metrics for Text Accuracy

Different from the WER, the semantics behind the transmitted sentence should be compared. Therefore, two new metrics are introduced to measure the performance of text recovery.

Bilingual evaluation understudy (BLEU) score is a widely used metric for evaluating the performance of machine translation systems. It quantifies the difference in n-grams between the transmitted and received sentences, where n-grams refer to contiguous sequences of n words. For instance, consider the sentence "machine translation is amazing,"

- 1-gram: "machine," "translation," "is," and "amazing";
- 2-grams: "machine translation," "translation is," and "is amazing";
- 3-grams: "machine translation is" and "translation is amazing."

The same rule applies to the rest.

For a reference sentence $\boldsymbol{s}$ of length l_s and a candidate sentence $\hat{\boldsymbol{s}}$ of length $l_{\hat{s}}$, the BLEU score is mathematically expressed as

$$\log \text{BLEU} = \min\left(1 - \frac{l_{\hat{s}}}{l_s}, 0\right) + \sum_{n=1}^{N} w_n \log p_n, \tag{3.18}$$

where w_n represents the weights assigned to n-grams, and p_n denotes the precision score computed from the frequency of matching n-grams.

The BLEU score spans from zero to one, with one signifying maximum similarity between the candidate and reference sentences. Notably, even human translations seldom attain a perfect score, as lexical variations (e.g., "I have a phone" vs. "I own a mobile") can alter the score without changing semantic meaning. To overcome this constraint, semantic similarity metrics are often used alongside BLEU to evaluate equivalence in meaning rather than just surface-level word overlap.

Sentence similarity: words can have different meanings depending on the context in which they are used. For example, the word "mouse" can refer to an animal in biology or a computer device in technology. Traditional methods, such as *word2vec* [9], struggle to recognize polysemy, which is the phenomenon where a single word can have multiple meanings. This challenge arises from the difficulty of representing a word with a numerical vector according to different contexts. To tackle this challenge, a new method measures semantic similarity between an original sentence, $\boldsymbol{s}$, and the recovered sentence, $\hat{\boldsymbol{s}}$, using the following cosine similarity formula.

$$\text{match}\,(\hat{\boldsymbol{s}}, \boldsymbol{s}) = \frac{f_{\Phi}\,(\boldsymbol{s}) \cdot f_{\Phi}(\hat{\boldsymbol{s}})^T}{\|\, f_{\Phi}\,(\boldsymbol{s})\, \|\, \|\, f_{\Phi}\,(\hat{\boldsymbol{s}})\, \|}, \tag{3.19}$$

where f_Φ denotes BERT [10], a large-scale pre-trained language model with billions of parameters that encodes semantic features of text. The similarity score in (3.19) spans from 0 to 1, quantifying how closely $\hat{\boldsymbol{s}}$ aligns with $\boldsymbol{s}$: a value of 1 signifies identical meaning, while 0 denotes no semantic overlap.

Unlike metrics such as BLEU, BERT leverages its training on vast corpora to generate context-aware semantic representations. For instance, the transmitted sentence $\boldsymbol{s}$ can be encoded into a semantic vector $\boldsymbol{c}$ via BERT, and the decoded sentence $\hat{\boldsymbol{s}}$ into a vector $\hat{\boldsymbol{c}}$. The cosine similarity between $\boldsymbol{c}$ and $\hat{\boldsymbol{c}}$ then serves as a robust measure of semantic equivalence between the two sentences, capturing nuanced contextual relationships beyond word matching.

3.4.2 Performance Metrics for Speech Quality

The effective reconstruction of speech signals is directly reflected in a satisfactory listening experience, characterized by the absence of latency and background noise in the restored speech. To this end, the PESQ [7], as recommended by ITU-T P.862, serves as a reliable performance metric for assessing speech signal quality across different scenarios, such as background noise and varying delays, scoring the quality on a scale from −0.5 to 4.5. In Figure 3.13, the PESQ evaluation model receives two speech inputs, including the reference speech and the degraded speech, and produces the score within a few milliseconds.

3.4.3 Performance Metrics for Image Quality

In the process of coding and transmission of images, various types of distortions may occur, including resolution degradation/blurring, frame stuttering, and block artifacts caused by quantization. These distortions result in degraded quality, necessitating scientific evaluation metrics to measure and perceive these changes and distortions, thereby improving user experience.

The evaluation of image quality is divided into subjective and objective metrics. Subjective metrics directly rely on user perception by calculating mean opinion score (MOS) or difference mean opinion score (DMOS) to reflect the

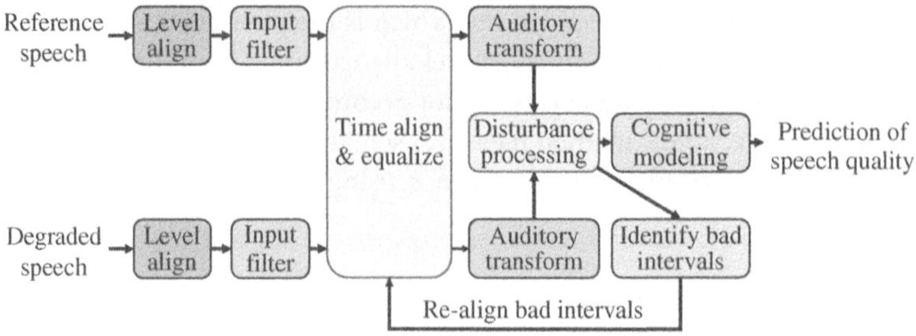

Figure 3.13 The structure of PESQ assessment model.

quality of images. While subjective metrics accurately reflect human perception, they are costly and time-consuming. Objective metrics calculate the differences between compressed images and their original counterparts, providing efficient and reproducible assessments that are widely used in algorithm development and optimization.

Image semantic similarity measures the degree of similarity between two images and is defined as

$$\eta(f(A), f(B)) = \| f(A) - f(B) \|_2^2, \tag{3.20}$$

where $f(\cdot)$ is the image embedding function that maps an image into Euclidean space. The embedding function $f(\cdot)$ is a critical part of semantic similarity measurements, while traditional metrics, such as PSNR and structural similarity index (SSIM), have certain limitations in capturing complex human perceptual features.

1) Traditional objective evaluation metrics: Traditional objective metrics are classified into two categories: error-based metrics and human visual system (HVS)-based metrics. Error-based metrics compute pixel-wise differences between distorted and original images, treating such differences as noise or distortion. HVS-based metrics incorporate models of human visual perception to mathematically describe how humans perceive structural differences, brightness, contrast, and other phenomena.

- MSE measures the squared difference between pixel intensities of the original and distorted images,

$$\text{MSE} = \frac{1}{N} \sum_{i=1}^{N} (x_i - y_i)^2, \tag{3.21}$$

- PSNR is a widely used metric for evaluating the quality of compressed images. It measures the ratio of the peak signal power to the noise power, expressed in decibels (dB). A higher PSNR value indicates better image quality, as it indicates a lower level of distortion in the compressed image. PSNR is calculated as follows,

$$\text{PSNR} = 10 \log_{10}\left(\frac{\text{MAX}^2}{\text{MSE}}\right), \tag{3.22}$$

where MAX is the maximum possible pixel value (e.g., 255 for 8-bit images).

- SSIM [11] is a perceptual metric that measures the similarity between two images based on luminance, contrast, and structure. It compares the luminance, contrast, and structure of the original and compressed images to assess the overall quality of the compression. SSIM values range from −1 to 1, with 1 indicating perfect similarity. SSIM is calculated as follows:

$$\text{SSIM}(x, y) = \frac{(2\mu_x\mu_y + C_1)(2\sigma_{xy} + C_2)}{(\mu_x^2 + \mu_y^2 + C_1)(\sigma_x^2 + \sigma_y^2 + C_2)}, \tag{3.23}$$

where μ_x and μ_y are the mean values of the original and distorted images, σ_x^2 and σ_y^2 are variances, and σ_{xy} is the covariance. Constants C_1 and C_2 stabilize the computation.

MS-SSIM [12] is an extension of SSIM that incorporates multi-scale information to better capture the perceptual quality of compressed images. It computes SSIM at multiple scales and combines the results to provide a more comprehensive assessment of image quality. MS-SSIM is calculated as follows:

$$\text{MS-SSIM}(x, y) = \frac{1}{L}\sum_{i=1}^{L} \text{SSIM}(x_i, y_i), \tag{3.24}$$

where L is the number of scales, and x_i and y_i are the images at scale i.

2) Modern objective evaluation metrics: Modern objective evaluation metrics leverage DL to extract high-level features from images, improving upon traditional methods in simulating human perception. These metrics use pre-trained DNNs to capture semantic and structural information in images.

- Inception score (IS) [13] evaluates the clarity and diversity of generated images,

$$\text{IS} = \exp\left(\mathbb{E}_{x\, p_g}\left[\text{D}_{\text{KL}}(p(y|x)||p(y))\right]\right), \tag{3.25}$$

where $p(y|x)$ is the conditional probability distribution of class y given image x, p_g is the distribution of the generative model, and D_{KL} is the KL-divergence. This metric measures two important aspects of the generated images. The first is the clarity and sharpness of the generated image, because if the image is blurry or of poor quality, it will be difficult to determine its category accurately, and the category probability distribution will be more chaotic, resulting in a lower IS. The second is the diversity of the generated images, because if the generated images are concentrated in a few categories, the edge probabilities will be more concentrated, and the KL-divergence value will also be affected so that IS can reflect the diversity of the generated images in terms of categories.

- The basic idea of Fréchet inception distance (FID) [14] is to map both the generated image and the real image into a high-dimensional feature space through a NN. In this high-dimensional feature space, the feature distribution of the real image and the feature distribution of the generated image can be approximated by a multivariate Gaussian distribution. FID is calculated as follows:

$$\text{FID} = \|\mu_r - \mu_g\|^2 + \text{Tr}(\Sigma_r + \Sigma_g - 2(\Sigma_r\Sigma_g)^{1/2}), \tag{3.26}$$

where μ_r and Σ_r are the mean and covariance of real images, and μ_g and Σ_g are the mean and covariance of generated images.

- Learned perceptual image Patch similarity (LPIPS) [15] is a perceptual metric that measures the similarity between two images based on features extracted from pre-trained DNNs. Its core assumption is that NNs pre-trained on large-scale image datasets have learned some important features of the HVS to perceive images, such as texture, color, structure, and so forth. In the feature space of these networks, the distance between image features can better reflect the image differences perceived by humans. LPIPS is calculated as the Euclidean distance between the feature representations of the original and compressed images, denoted as

$$\text{LPIPS}(x, y) = \sum_{l=1}^{L} w_l \cdot \|\phi_l(x) - \phi_l(y)\|_2^2, \tag{3.27}$$

where $\phi_l(x)$ represents the feature extracted from image x at layer l, and w_l is a channel weight.

- Visual semantic embedding [16, 17] evaluates semantic similarity by mapping image features to low-dimensional semantic spaces, including the VTransE network that maps object and relationship visual features to low-dimensional spaces. It has been applied in scene graph generation, image captioning, and retrieval.

References

1 H. Xie, Z. Qin, G. Y. Li, and B.-H. Juang, "Deep learning enabled semantic communication systems," *IEEE Trans. Signal Process.*, vol. 69, pp. 2663–2675, 2021.

2 M. I. Belghazi, A. Baratin, S. Rajeshwar, et al., "Mutual information neural estimation," in *Proc. Int. Conf. Mach. Learning (ICML)*. PMLR, 2018, pp. 531–540.
3 Z. Weng and Z. Qin, "Semantic communication systems for speech transmission," *IEEE J. Sel. Areas Commun.*, vol. 39, no. 8, pp. 2434–2444, 2021.
4 J. Hu, L. Shen, S. Albanie, et al., "Squeeze-and-excitation networks," *IEEE Trans. Pattern Anal. Mach. Intell.*, vol. 42, no. 8, pp. 2011–2023, 2020.
5 H. S. Black and J. O. Edson, "Pulse code modulation," *Trans. AIEE*, vol. 66, no. 1, pp. 895–899, 1947.
6 C. Berrou and A. Glavieux, "Near optimum error correcting coding and decoding: turbo-codes," *IEEE Trans. Commun.*, vol. 44, no. 10, pp. 1261–1271, 1996.
7 A. W. Rix, J. G. Beerends, M. P. Hollier, and A. P. Hekstra, "Perceptual evaluation of speech quality (PESQ)-a new method for speech quality assessment of telephone networks and codecs," in *Proc. IEEE Int. Conf. Acoust., Speech, Signal Process. (ICASSP)*, Salt Lake City, UT, USA, 2001, pp. 749–752.
8 H. Xie, Z. Qin, Z. Han, and K. B. Letaief, "Hybrid digital-analog semantic communications," *arXiv preprint arXiv:2405.12580*, 2024.
9 T. Mikolov, K. Chen, G. Corrado, and J. Dean, "Efficient estimation of word representations in vector space," in *Proc. Int. Conf. Learn. Represent. (ICLR)*, Scottsdale, AZ, USA, 2013.
10 J. Devlin, M.-W. Chang, K. Lee, and K. Toutanova, "BERT: pre-training of deep bidirectional transformers for language understanding," in *Proc. Conf. N. Am. Chapter Assoc. Comput. Linguistics: hum. Lang. Technol. (NAACL HLT)*, vol. 1, Minneapolis, MN, USA, 2019, pp. 4171–4186.
11 Z. Wang, A. C. Bovik, H. R. Sheikh, and E. P. Simoncelli, "Image quality assessment: from error visibility to structural similarity," *IEEE Trans. Image Process.*, vol. 13, no. 4, pp. 600–612, 2004.
12 Z. Wang, E. P. Simoncelli, and A. C. Bovik, "Multiscale structural similarity for image quality assessment," in *Proc. Thirty-Seventh Asilomar Conf. Signals, Syst. Comput.*, vol. 2, 2003, pp. 1398–1402.
13 T. Salimans, I. J. Goodfellow, W. Zaremba, et al., "Improved techniques for training gans," in *Proc. Advances Neural Inf. Processing Syst. (NIPS)*, 5–10, 2016, pp. 2226–2234.
14 M. Heusel, H. Ramsauer, T. Unterthiner, et al., "Gans trained by a two time-scale update rule converge to a local nash equilibrium," in *Proc. Advances Neural Inf. Processing Syst. (NIPS)*, 4–9, 2017, pp. 6626–6637.
15 R. Zhang, P. Isola, A. A. Efros, et al., "The unreasonable effectiveness of deep features as a perceptual metric," in *Proc. IEEE Comput. Soc. Conf. Comput. Vis. Pattern Recognit. (CVPR)*, 2018, pp. 586–595.

16 H. Zhang, Z. Kyaw, S.-F. Chang, and T.-S. Chua, "Visual translation embedding network for visual relation detection," in *Proc. IEEE Conf. Comput. Vis. Pattern Recognit. (CVPR),* Honolulu, Hawaii, 2017, pp. 5532–5540.

17 J. Zhang, Y. Kalantidis, M. Rohrbach, et al., "Large-scale visual relationship understanding," in *Proc. AAAI Conf. Artif. Intell. (AAAI),* Honolulu, Hawaii, 2019, pp. 9185–9194.

4

Task-oriented Semantic Communications

This chapter introduces task-oriented semantic communications, which aim to directly execute the intelligent tasks, for example, the machine translation and speech-to-text. This approach can significantly enhance the efficiency and effectiveness of communication systems by extracting only the task-related information. Due to directly optimizing the system with the tasks, these systems can further reduce the amount of data that needs to be transmitted and improve the performance of the task. The following sections will delve into the specific frameworks and methodologies for different task-oriented semantic communication systems, including single-modal and multimodal systems.

Task-oriented semantic communications prioritize the significance of information as semantics. As illustrated in Figure 4.1, these communication systems are designed with specific transmission goals, such as intelligent tasks, for example, image classification, machine translation, robot control, and more. By focusing on these goals, the system filters out goal-relevant information while discarding irrelevant data. The deep learning (DL)-based systems share similarities with the DL-based semantic communication systems discussed in Chapter 3, as they primarily rely on deep neural networks (DNNs). However, the key distinction lies in their purpose—task-oriented systems directly execute intelligent tasks rather than reconstructing data.

4.1 Single-modal Task-oriented Semantic Communications

This section will introduce the single-modal task-oriented semantic communications, in which only one modality is considered for the task execution.

Deep Learning Enabled Semantic Communications, First Edition. Zhijin Qin et al.

Published 2026 by John Wiley & Sons, Inc.

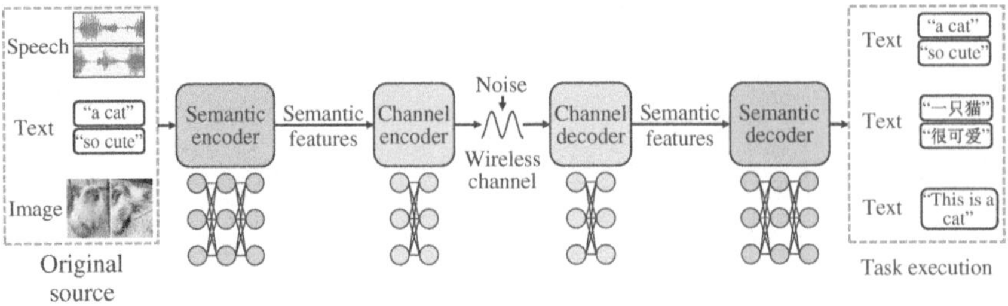

Figure 4.1 The structure of joint semantic-channel coding for task execution.

4.1.1 Semantic Communications for Machine Translation

4.1.1.1 System Model

Consider a multiple-input multiple-output (MIMO) communication system comprising K single-antenna users and a base station equipped with M antennas. The system consists of a semantic transmitter and a semantic receiver.

4.1.1.1.1 Semantic Transmitter

Given the source language sentence of k-th user (e.g., English), $\boldsymbol{s}_k$. The semantic information is initially extracted by

$$\boldsymbol{z}_k = S_{\text{MT}}(\boldsymbol{s}_k; \boldsymbol{\alpha}_k), \tag{4.1}$$

where $\boldsymbol{z}_k \in \mathbb{R}^{L_S \times 1}$ represents the semantic feature vector of length L_S. The function $S_{\text{MT}}(\cdot; \boldsymbol{\alpha}_k)$ denotes the semantic encoder for the k-th user, parameterized by the learnable parameters $\boldsymbol{\alpha}_k$. Then, the transmitted symbols of the k-th user are subsequently obtained by

$$\boldsymbol{x}_k = C_{\text{MT}}(\boldsymbol{z}_k; \boldsymbol{\beta}_k), \tag{4.2}$$

where $\boldsymbol{x}_k \in \mathbb{C}^{L_C \times 1}$ represents the transmitted complex signal of length $L_C < L_S$. The function $C_{\text{MT}}(\cdot; \boldsymbol{\beta}_k)$ denotes the channel encoder for the k-th user. The neural channel encoder is designed to compress semantic information, thereby reducing the number of transmitted symbols and enhancing robustness to channel variations.

4.1.1.1.2 Semantic Receiver

After transmitting over MIMO physical channels, the received signal can be expressed as

$$\boldsymbol{Y} = \boldsymbol{H}\boldsymbol{X} + \boldsymbol{N}, \tag{4.3}$$

where $\boldsymbol{X}^T = [\boldsymbol{x}_1, \boldsymbol{x}_2, \cdots, \boldsymbol{x}_K] \in \mathbb{C}^{L_C \times K}$ denotes the transmitted symbols from all K users, and $\boldsymbol{H} = [\boldsymbol{h}_1, \boldsymbol{h}_2, \ldots, \boldsymbol{h}_K] \in \mathbb{C}^{M \times K}$ represents the channel matrix between the base station and the users. $\boldsymbol{N} \in \mathbb{C}^{M \times L_C}$ denotes the circular symmetric Gaussian noise, with elements that are i.i.d. with zero mean and variance σ_n^2.

Subsequently, the transmitted signals are recovered using the L-MMSE detector,

$$\hat{\boldsymbol{X}} = \hat{\boldsymbol{H}}^{\mathrm{H}}\left(\hat{\boldsymbol{H}}\hat{\boldsymbol{H}}^{\mathrm{H}} + \sigma_n^2 \mathbf{I}\right)^{-1} \mathbf{Y}, \tag{4.4}$$

where $\hat{\boldsymbol{X}}^T = [\hat{\boldsymbol{x}}_1; \hat{\boldsymbol{x}}_2; \cdots; \hat{\boldsymbol{x}}_K] \in \mathbb{C}^{L_C \times K}$ is the recovered transmitted signals, $\hat{\boldsymbol{H}} = \boldsymbol{H} + \Delta\boldsymbol{H}$ is the estimated channel state information (CSI), in which $\Delta\boldsymbol{H}$ is the estimation error with $\Delta\boldsymbol{H} \in \mathcal{CN}(0, \sigma_e^2)$.

The semantic feature of the k-th user, $\hat{\boldsymbol{z}}_k \in \mathbb{R}^{L_S \times 1}$, is recovered by the JSC decoder as

$$\hat{\boldsymbol{z}}_k = C_{\mathrm{MT}}^{-1}(\hat{\boldsymbol{x}}_k; \boldsymbol{\gamma}_k), \tag{4.5}$$

where $C_{\mathrm{MT}}^{-1}(\hat{\boldsymbol{x}}_k; \boldsymbol{\gamma}_k)$[1] represents the channel decoder for the k-th user, parameterized by $\boldsymbol{\gamma}_k$. The purpose of the channel decoder is to reduce the impact of inter-user interference and channel distortion while decompressing the semantic information.

For single-modal semantic transmission, each user independently executes various tasks. Consequently, the following is how the recovered semantic information is applied to the k-th user's task,

$$\boldsymbol{p}_k = S_{\mathrm{MT}}^{-1}(\hat{\boldsymbol{z}}_k; \boldsymbol{\varphi}_k), \tag{4.6}$$

where $\boldsymbol{p}_k$ is the translated sentence for the machine learning task. $S_{\mathrm{MT}}^{-1}(\cdot; \boldsymbol{\varphi}_k)$ is semantic decoder for the k-th user with learning parameters $\boldsymbol{\varphi}_k$.

4.1.1.2 Model Description

The system's architecture is illustrated in Figure 4.2. At the transmitter end, a text semantic encoder and a text channel encoder work sequentially to extract and compress the textual semantic information. The semantic encoder employs multiple Transformer encoder layers, whereas the channel encoder, as shown in Figure 4.2, incorporates a series of dense layers. Transformers are chosen over recurrent neural networks (RNNs) owing to their enhanced capacity to model contextual relationships across sentences and their considerably faster training efficiency. On the receiver side, the text channel decoder aims to reconstruct semantic information from corrupted signals. Subsequently, the semantic decoder—equipped with multiple Transformer decoder layers—produces the translated output using the restored semantic data.

---

1 To simplify the notation, $\cdot^{-1}$ is used here to denote the decoder.

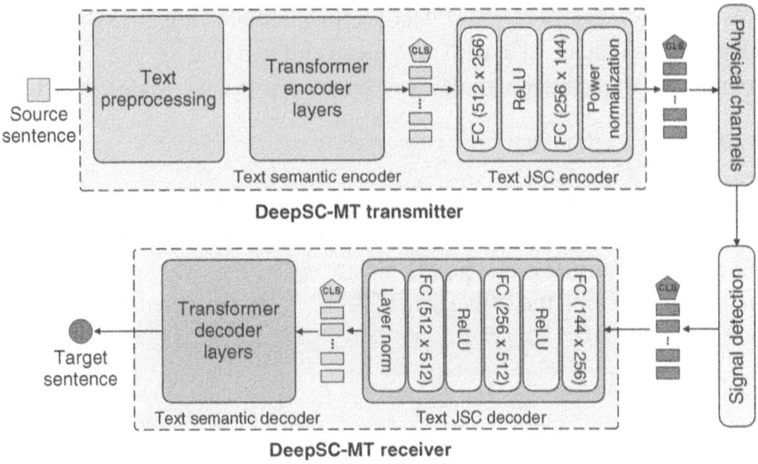

Figure 4.2 The network structure of semantic communication for machine translation task.

4.1.1.3 Model Training

Assume $\mathcal{D}_k = \{(\boldsymbol{s}_{k,j}, \boldsymbol{p}_{k,j})\}_{j=1}^{D}$ with size D as the training text dataset for the k-th user, where $\boldsymbol{s}_{k,j}$ and $\boldsymbol{p}_{k,j}$ are the j-th sentence in the source language and the translated sentence in the target language, respectively. $\boldsymbol{s}_{k,j}^{\mathcal{T}}[n]$ and $\boldsymbol{p}_{k,j}^{\mathcal{T}}[n]$ represent the n-th word in sentence $\boldsymbol{s}_{k,j}^{\mathcal{T}}$ and $\boldsymbol{p}_{k,j}^{\mathcal{T}}$, respectively.

The training process consists of three phases. The first phase is trained using the cross-entropy (CE) loss function. This training enables the model to learn the distribution of words in the target language, thereby facilitating the conversion of meaning into the target sentence. The CE loss function is defined as

$$\mathcal{L}_{\mathrm{MT}} = \mathbb{E}\left[-\sum_{n} P(\boldsymbol{p}_{k,j}^{\mathcal{T}}[n]) \log\left(P(\hat{\boldsymbol{p}}_{k,j}^{\mathcal{T}}[n])\right)\right], \tag{4.7}$$

which measures the discrepancy between the predicted word probabilities $P(\hat{\boldsymbol{p}}_{k,j}^{\mathcal{T}}[n])$ and the ground-truth distribution $P(\boldsymbol{p}_{k,j}^{\mathcal{T}}[n])$. This phase enables the model to capture syntactic, phrasal, and semantic patterns in the target language.

The second training phase is achieved using the mean squared error (MSE) loss function, defined as

$$\mathcal{L}_{\mathrm{MSE}} = \mathbb{E}\left[\left\|\hat{\boldsymbol{z}}_{k,j}^{\mathcal{T}} - \boldsymbol{z}_{k,j}^{\mathcal{T}}\right\|_2^2\right], \tag{4.8}$$

where $\boldsymbol{z}_{k,j}^{\mathcal{T}}$ is the transmitted semantic embedding and $\hat{\boldsymbol{z}}_{k,j}^{\mathcal{T}}$ is its reconstructed version at the receiver. This phase aims to compress semantic information and mitigate channel distortions and multi-user interference.

In the final phase, the entire system is fine-tuned jointly using the CE loss to reduce semantic discrepancies between $\hat{z}_{k,j}^{T}$ and $z_{k,j}^{T}$, thereby minimizing error propagation across modules.

4.1.1.4 Numerical Results

Figure 4.3 illustrates the bilingual evaluation understudy (BLEU) score comparison for English-to-Chinese translation across varying channel conditions. The results indicate that the SemCom system achieves a higher BLEU score than the UTF-8-Turbo method combined with Quadrature Phase Shift Keying (QPSK) modulation, particularly in low signal-to-noise ratio (SNR) environments over additive white Gaussian noise (AWGN) channels, as well as across all SNR levels in fading channels. While both systems experience BLEU score declines due to imperfect CSI, the semantic communication system exhibits greater resilience, consistently surpassing the baseline even under suboptimal CSI conditions. This demonstrates the robustness of the SemCom system in handling channel distortions and multi-user interference, thereby enhancing the translation performance in challenging communication scenarios.

4.1.2 Semantic Communications for Speech Recognition and Synthesis

With the ubiquitous deployment of intelligent devices in the post-Shannon communication era, substantial data transmission has become essential to enable massive connectivity among these devices. Additionally, the implementation of downstream tasks at the receiver raises the need for cross-modal and multimodal intelligent transmission. However, the scarce spectrum resources result in a significant challenge in supporting extensive tasks, especially for speech-related tasks that demand enormous data transmission. To combat this, the source message can be condensed at the semantic level to remove information irrelevant to the task while extracting only the task-associated semantic features. This is because the data produced at the receiver is not required to align precisely with the source data at the binary level and typically diverges from it. In this section, task-oriented semantic communication systems for speech transmission are presented, and a specific example of DeepSC-ST [1] is introduced, which executes speech recognition and speech synthesis tasks by transmitting low-dimensional semantic features.

The speech signals mainly consist of acoustic characteristics and textual information. In some communication scenarios, the receiver may only require the textual information. Inspired by this, the input speech is initially compressed to omit the acoustic characteristics, such as speaker information and speech delay, prior to transmission using a complex neural network (NN). By doing so, the large semantic redundancy within the source speech is removed, and the text-related

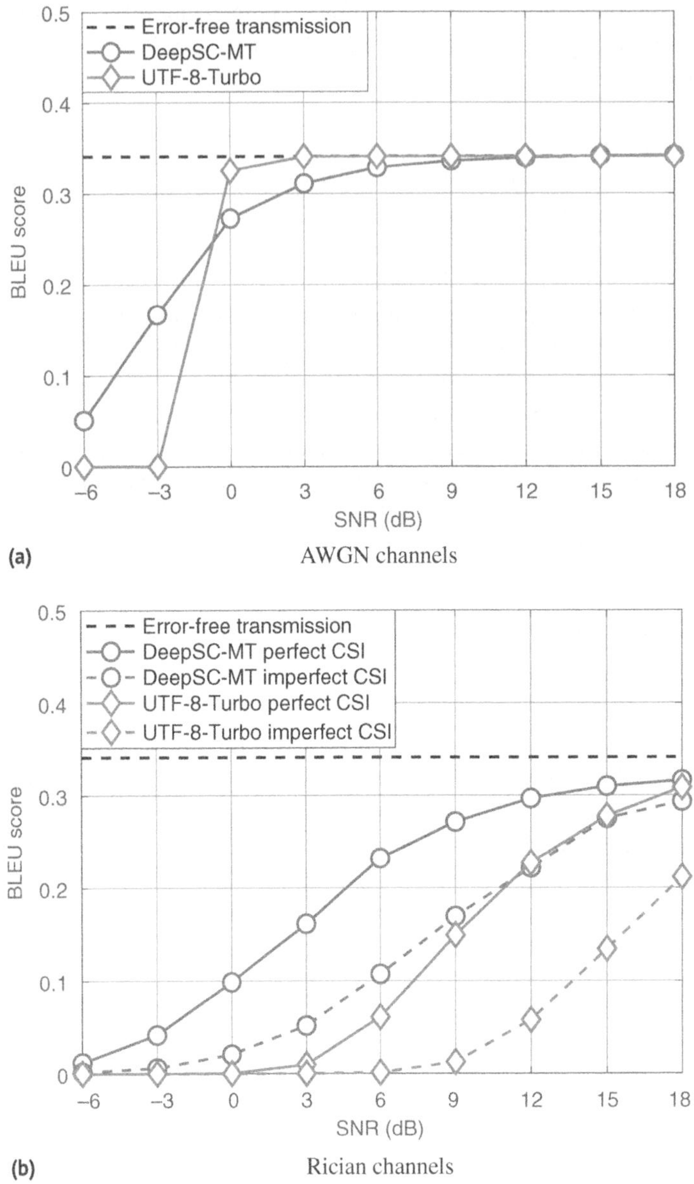

Figure 4.3 BLEU score comparison between DeepSC-MT and UTF-8-Turbo with QPSK under (a) AWGN channels and (b) Rician channels.

semantic features are extracted, which significantly reduces the amount of transmission data because acoustic characteristics occupy a large proportion of the speech data. Furthermore, thanks to the success of text-to-speech conversion techniques, the source speech can be recovered at the receiver based on the obtained textual information using a speech synthesis module. The model structure of the task-oriented semantic communications system, DeepSC-ST, is shown in Figure 4.4. From the figure, the DeepSC-ST can serve users requesting both text and speech. It is worth mentioning that the normalization and framing operations are performed on the input speech before feeding it into the NN-enabled joint semantic-channel coding.

4.1.2.1 Joint Semantic-channel Coding

The objective of the semantic encoder is to compress the input speech as much as possible while preserving the fidelity of the textual semantics to the fullest extent. Toward this end, multiple 1D CNNNs are leveraged in the DeepSC-ST to eliminate acoustic characteristics completely. In addition, the gated recurrent unit (GRU)-based bidirectional recurrent neural network (BRNN) [2] is adopted to obtain textual features due to its strong capability to capture the contextual dependencies, especially for lengthy sentences. As shown in Figure 4.5, the hidden feature, $\boldsymbol{F}_1$, is a floating-point value-based matrix that is passed through the BRNN to extract the semantic feature, $\boldsymbol{F}$. Note that $\boldsymbol{F}$ is a probability matrix, with each probability vector representing the probabilities associated with mappings to tokens in the predefined vocabulary.

The semantic features, $\boldsymbol{F}$, are fed into a dense layer-enabled channel encoder, and the encoded symbols, $\boldsymbol{X}$, are transmitted over the wireless channel. At the receiver, the channel decoder is responsible for recovering accurate semantic features, $\hat{\boldsymbol{F}}$. The objective of DeepSC-ST for the speech recognition task is to retrieve the text from the input speech after the wireless transmission process, which involves maximizing the posterior probability, $p(\boldsymbol{T}|\ \boldsymbol{S})$, where $\boldsymbol{T}$ is the

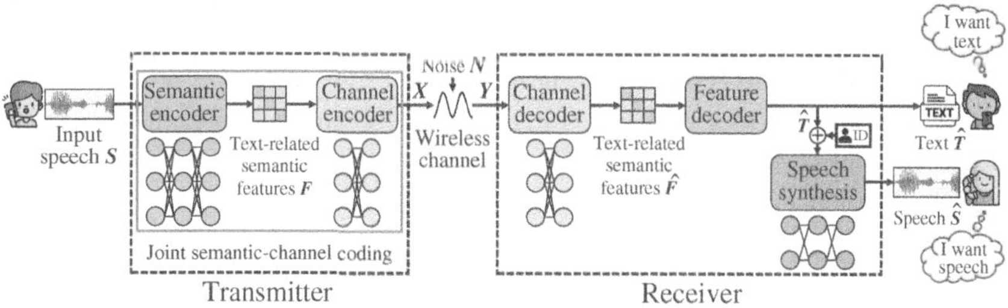

Figure 4.4 The model structure of the DL-enabled semantic communication system for speech recognition and speech synthesis.

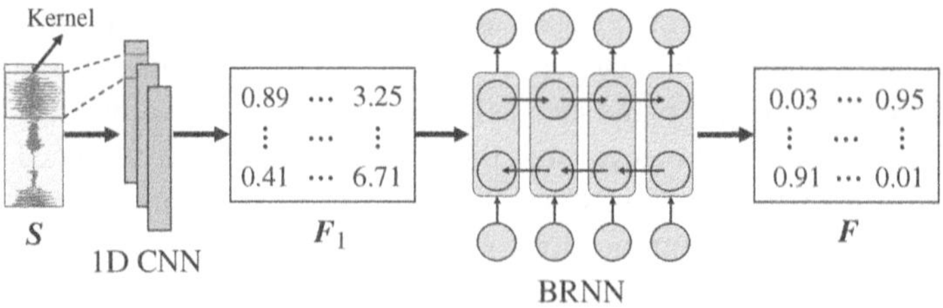

Figure 4.5 1D CNN and BRNN-enabled semantic encoder.

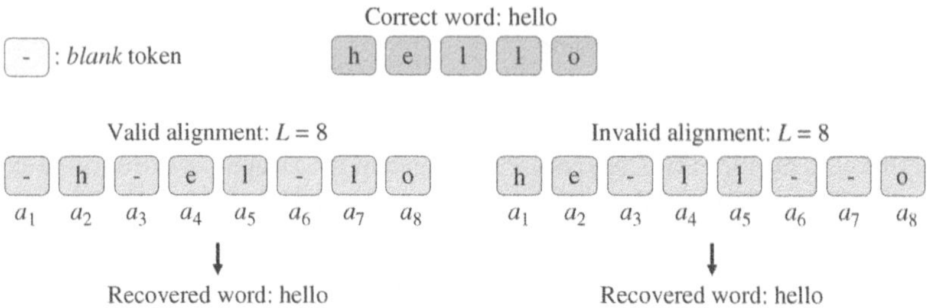

Figure 4.6 An example of mapping the valid alignment to text.

accurate text. According to the connectionist temporal classification (CTC) [3], $p(\boldsymbol{T}|\ \boldsymbol{S})$ can be expressed as

$$p(\boldsymbol{T}|\ \boldsymbol{S}) = \sum_{A \in \mathfrak{A}(\boldsymbol{S},\ \boldsymbol{T})} \left(\prod_{l=1}^{L} \hat{P}_l (a_l|\ \boldsymbol{S}) \right), \tag{4.9}$$

where $\hat{P}_l$ and a_l are a probability and a token in the given vocabulary, respectively. $\mathfrak{A}(\boldsymbol{S},\ \boldsymbol{T})$ indicates the set of all possible valid alignments between $\boldsymbol{T}$ and $\hat{\boldsymbol{T}}$. For instance, consider $\boldsymbol{T} = [\text{h, e, l, l, o}]$, the valid alignments include [*blank*, h, *blank*, e, l, *blank*, l, o] and [h, *blank*, e, l, *blank*, *blank*, l, o] because the *blank* token is removed to produce $\hat{\boldsymbol{T}}$. An example of obtaining the text from an alignment is shown in Figure 4.6.

Therefore, the CTC loss is used to update the NN parameters of the semantic encoder, the channel encoder, and the channel decoder, θ, by maximizing the $p(\boldsymbol{T}|\ \boldsymbol{S})$, denoted as

$$\mathcal{L}_{\text{CTC}}(\boldsymbol{S},\ \boldsymbol{T};\ \theta) = -\ln \left(\sum_{A \in \mathfrak{A}(\boldsymbol{S},\ \boldsymbol{T})} \left(\prod_{l=1}^{L} \hat{P}_l (a_l|\ \boldsymbol{S}, \theta) \right) \right). \tag{4.10}$$

4.1.2.2 Greedy Decoder

As mentioned earlier, the channel decoder generates accurate semantic features, $\widehat{\boldsymbol{F}}$, once all NNs have been adequately trained. According to $\widehat{\boldsymbol{F}}$, the text is obtained by implementing the greedy decoding, as shown in Figure 4.7. In the decoding process, at each step, the token corresponding to the highest probability in the vector $\widehat{\boldsymbol{P}}_l$ is chosen and mapped to its respective token in vocabulary $\overline{\boldsymbol{T}}$. As a result, the final text, $\widehat{\boldsymbol{T}}$, is produced by concatenating all the selected tokens. It is important to note that the steps of selecting the maximum probability in $\widehat{\boldsymbol{P}}_l$ and mapping it to the corresponding token in $\widehat{\boldsymbol{T}}$ are non-differentiable. This conflicts with the requirement for a differentiable loss function when designing the NN. Therefore, the greedy decoder cannot be substituted by the NN.

4.1.2.3 Speech Synthesis Module

By incorporating a flexible task mechanism, the information generated at the receiver extends beyond the text sequence to include high-quality speech, which enables users to access both acoustic characteristics and textual information, thereby enhancing the versatility of DeepSC-ST and offering greater flexibility in its application. Inspired by this, the recovered text, $\widehat{\boldsymbol{T}}$, is input into the NN-enabled speech synthesis module and provides speech, $\hat{\boldsymbol{S}}$. The speech synthesis module is trained independently of the speech recognition task and can be excluded in communication scenarios where the requested information is limited to text only. Moreover, to guarantee that the synthesized speech resembles the source speaker, the speaker identity (ID) information is treated as additional information during synthesis, and it is pre-registered on the user device or stored at the server where it can be downloaded.

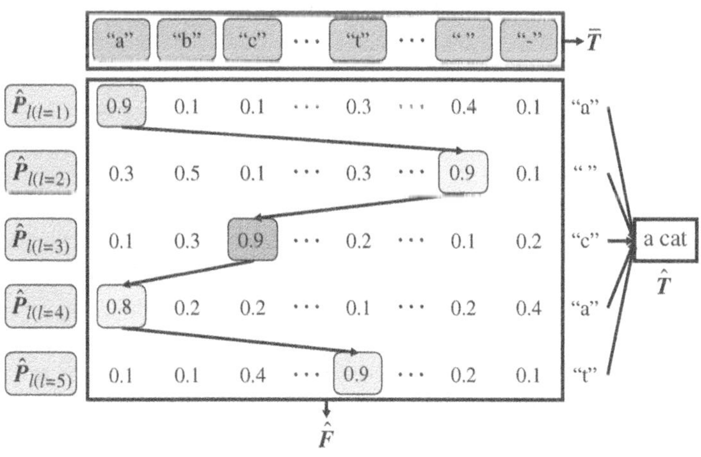

Figure 4.7 An example of obtaining text by the greedy decoder.

4.1.2.4 Performance Metrics

4.1.2.4.1 Speech Recognition Task

The textual semantic similarity between the recovered and accurate text is assessed by evaluating how readable and comprehensible the recovered text is. It is evident that text containing fewer errors, such as incorrect characters or words, exhibits higher intelligibility. Hence, the word error rate (WER) is considered an effective metric to measure the semantic integrity of the recognized text, denoted as

$$\mathrm{WER} = \frac{S_W + D_W + I_W}{N_W}, \tag{4.11}$$

where N_W represents the number of words in $\boldsymbol{t}$. S_W, D_W, and I_W are substitution, deletion, and insertion operations in $\hat{\boldsymbol{t}}$.

The threshold for an acceptable WER score is influenced by both the language being processed and individual subjective factors. In practice, the sentence is usually readable when WER is lower than around 0.15.

4.1.2.4.2 Speech Synthesis Task

The speech synthesis task commits to accurately reconstructing the clear speech. Due to the challenges associated with directly scoring speech quality through multiple human evaluators, an alternative method involves comparing it to real speech using an assessment model. In this context, the Fréchet Deep Speech Distance (FDSD) [4] is employed as a quantitative metric to measure the distributional similarity between the synthesized and real speech by calculating their Fréchet distance. Note that a higher FDSD score corresponds to lower speech quality. According to the real speech sequence, $\boldsymbol{S}$, and the synthesized speech sequence, $\hat{\boldsymbol{S}}$, an assessment model extracts the hidden features of $\boldsymbol{S}$ and $\hat{\boldsymbol{S}}$ and obtains $\boldsymbol{D} \in \mathbb{R}^{U\times V}$ and $\hat{\boldsymbol{D}} \in \mathbb{R}^{\hat{U}\times V}$,respectively. Then, the FDSD is calculated as

$$\mathrm{FDSD}^2 = \|\boldsymbol{\mu}_D - \boldsymbol{\mu}_{\hat{D}}\|^2 + \mathrm{Tr}\left[\boldsymbol{\Sigma}_D + \boldsymbol{\Sigma}_{\hat{D}} - (\boldsymbol{\Sigma}_D\boldsymbol{\Sigma}_{\hat{D}})^{\frac{1}{2}}\right], \tag{4.12}$$

where $\boldsymbol{\mu}_D$ and $\boldsymbol{\mu}_{\hat{D}}$ denotes the means of $\boldsymbol{D}$ and $\hat{\boldsymbol{D}}$, respectively, $\boldsymbol{\Sigma}_D$ and $\boldsymbol{\Sigma}_{\hat{D}}$ are the relative covariance matrices.

To verify the superiority of DeepSC-ST, four benchmarks are provided for performance comparison, as shown in Figure 4.8. The first benchmark is a conventional speech transmission system that employs the AMR-WB algorithm [5] for source coding, the polar code [6] for channel coding, and 64-QAM for modulation. The speech recognition module, wav2vec 2.0 [7], is deployed at the receiver to obtain the text from the reconstructed speech. The second benchmark generates the text using wav2vec 2.0 at the transmitter and considers the conventional communication paradigm for text transmission.

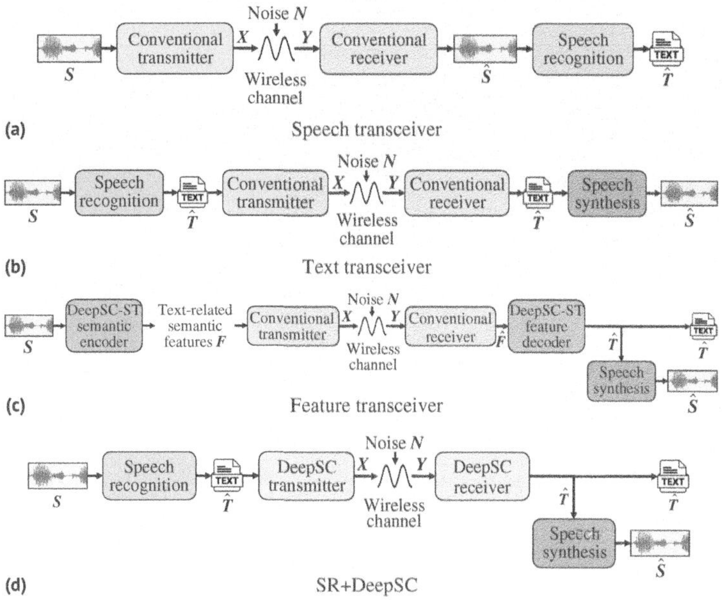

Figure 4.8 The model structure of different benchmarks: (a) speech transceiver, (b) text transceiver, (c) feature transceiver, and (d) SR+DeepSC.

The third benchmark transmits the extracted text-related semantic features produced by the semantic encoder of DeepSC-ST. In addition, the IEEE 754 floating-point arithmetic [8] module, the polar code, and the 64 QAM modulation scheme are adopted for feature transmission. The fourth benchmark is a hybrid modeling system that combines wav2vec 2.0 with the DeepSC [9] model, named SR+DeepSC.

The WER results of different methods are presented in Figure 4.9. As shown in the figure, the DeepSC-ST outperforms both the speech transceiver and text transceiver, achieving a lower WER under various channel conditions, particularly when the SNR is below 8 dB. Additionally, DeepSC-ST demonstrates consistently low WER on average against channel variations, whereas conventional systems and SR+DeepSC yield higher WER at low SNR. The simulation results indicate that DeepSC-ST significantly enhances the recovery of text at the receiver, offering superior performance in complex communication scenarios compared to both conventional and the existing semantic communication system, especially in the low SNR regime.

The FDSD results for DeepSC-ST and four benchmark models are illustrated in Figure 4.10, where the ground truth is derived from the FDSD scores obtained

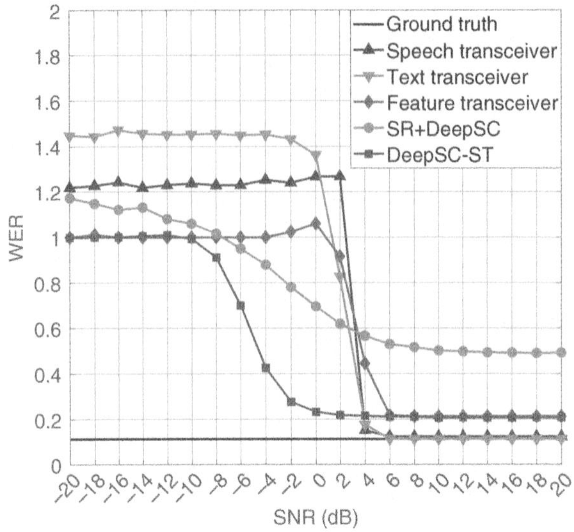

(a) AWGN channels

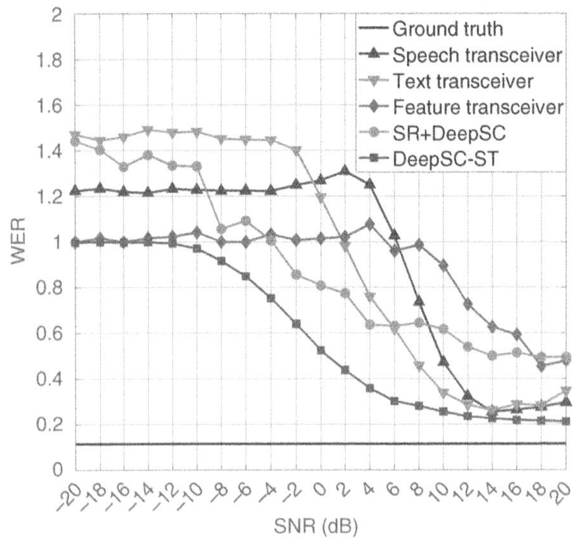

(b) Rayleigh channels

Figure 4.9 WER versus SNR for the speech transceiver, the text transceiver, the feature transceiver, the SR+DeepSC, and the DeepSC-ST: (a) AWGN channels and (b) Rayleigh channels.

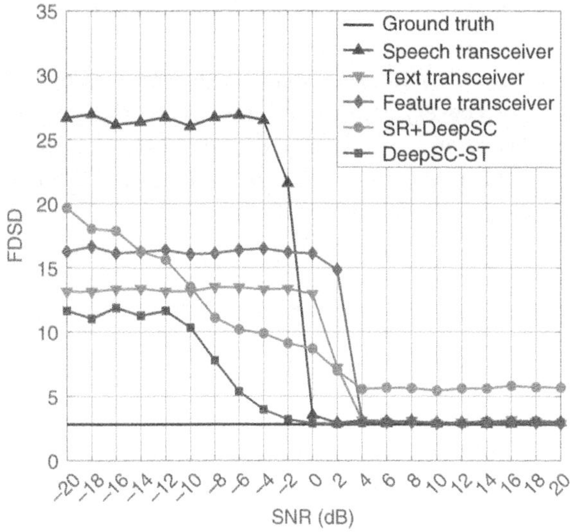

(a) AWGN channels

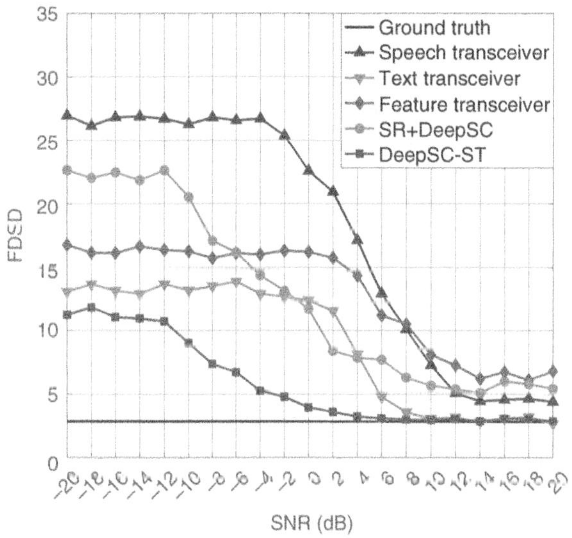

(b) Rayleigh channels

Figure 4.10 FDSD versus SNR for the speech transceiver, the text transceiver, the feature transceiver, the SR+DeepSC, and the DeepSC-ST: (a) AWGN channels and (b) Rayleigh channels.

by directly passing the plain text through the speech synthesis module, Tacotron 2 [10]. From the figure, DeepSC-ST achieves lower FDSD scores than the SR+DeepSC model across all tested channel conditions. Furthermore, DeepSC-ST demonstrates superior speech recovery compared to the speech transceiver, feature transceiver, and text transceiver across a range of SNRs from −12 dB to 2 dB under AWGN channels, and from nearly −12 dB to 10 dB under Rayleigh channels. Notably, when the SNR drops below approximately 4 dB under AWGN channels, the received speech signals in the speech transceiver become ineffective, whereas the synthesized waveform from the DeepSC-ST remains intelligible for SNRs greater than −10 dB, highlighting the robustness of the DeepSC-ST in low SNR environments.

4.1.2.5 Software Demonstration

A software demonstration with a user interface has been developed based on the DeepSC-ST, as shown in Figure 4.11.[2] As shown in the figure, users can either

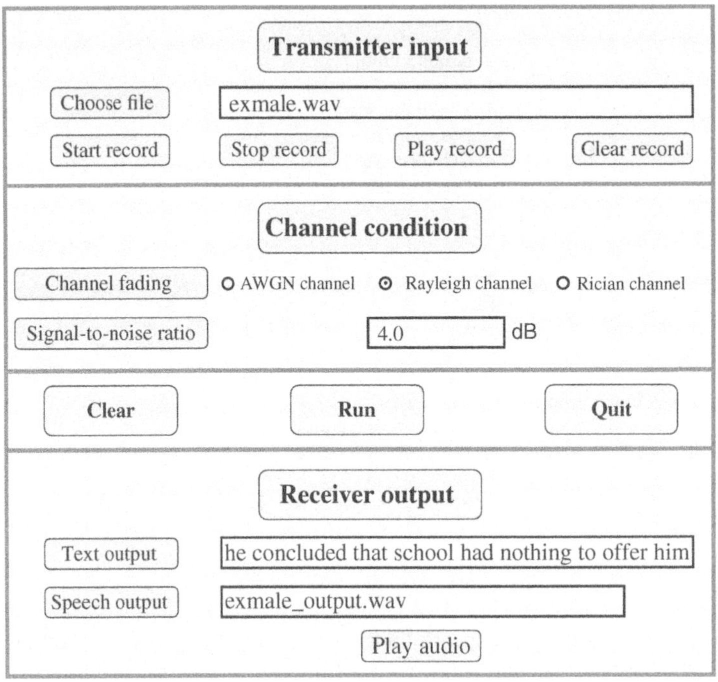

Figure 4.11 Software demonstration of the DeepSC-ST.

2 The software demonstration is available at https://github.com/Zhenzi-Weng/DeepSC-ST_demonstration.

select a local *.wav* file or record speech input. Additionally, the demonstration enables users to choose from three different fading channels and input an arbitrary SNR value. After running a pre-trained DeepSC-ST model, the recognized text and synthesized speech are generated as outputs.

4.2 Multimodal Task-oriented Semantic Communications

Multimodal data comes from multiple sources and can be combined to create a more complete understanding. It can include text, audio, video, images, and sensor data. In a multi-user transmission scenario, different users usually transmit different types of sources and perform various tasks at the receiver, which are typically multimodal tasks over the air. Next, a multimodal semantic communication system is introduced.

4.2.1 Semantic Communication Systems for Visual Question Answering

A multimodal MIMO communication scenario is considered, where single-antenna users transmit image and text information to the base station, respectively. As shown in Figure 4.12, the multimodal system consists of an image transmitter, a text transmitter, and a receiver.

4.2.1.1 Image Transmitter

The image transmitter, illustrated in Figure 4.12, consists of two core components: a semantic encoder and a channel encoder. The encoded semantic information is formalized as

$$z_I = S_I(s_I; \alpha_I), \tag{4.13}$$

where z_I represents the semantic features. Here, s_I denotes the input image, and α_I is the trainable parameters of the semantic encoder, S_I.

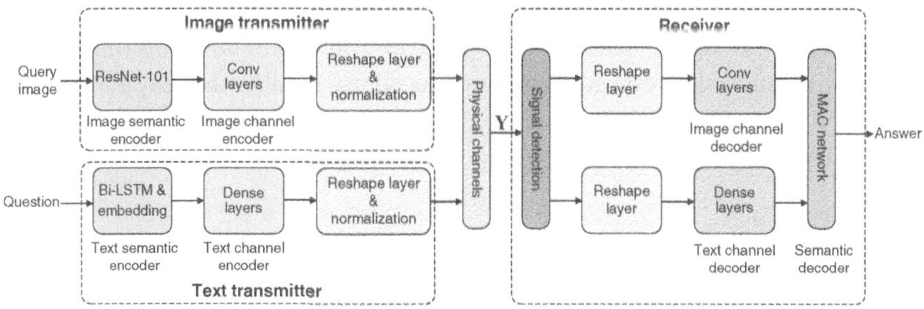

Figure 4.12 The structure of MU-DeepSC.

After semantic encoding, the extracted semantic information is mapped directly to transmitted symbols via an image channel encoder. The transmitted symbols can thus be mathematically represented as

$$\boldsymbol{x}_I = C_I\,(\boldsymbol{z}_I; \boldsymbol{\beta}_I), \tag{4.14}$$

where $\boldsymbol{x}_I$ corresponds to the compressed representation of the image's semantic information. $\boldsymbol{\beta}_I$ represents the trainable parameters of the channel encoder, C_I.

In the aforementioned process, $\boldsymbol{x}_I$ is represented as a real-valued signal, which is not directly suitable for transmission over physical channels. Consequently, $\boldsymbol{x}_I$ should be transformed into a complex-valued signal and normalized the power by $\boldsymbol{x}_I/\mathbb{E}(\|\boldsymbol{x}_I\|_2)$.

4.2.1.2 Text Transmitter

The text transmitter first utilizes the text semantic encoder to extract the semantic representations of the input sentence, which is formulated as

$$\boldsymbol{z}_T = S_T\,(\boldsymbol{s}_T; \boldsymbol{\alpha}_T), \tag{4.15}$$

where $\boldsymbol{z}_T$ is the text semantic feature, $\boldsymbol{s}_T$ denotes the input sentence, and $\boldsymbol{\alpha}_T$ indicates the trainable parameters of text semantic encoder S_T. Subsequently, the text is compressed and mapped to the transmitted text symbols via the channel encoder, which is given by

$$\boldsymbol{x}_T = C_T\,(\boldsymbol{z}_T; \boldsymbol{\beta}_T), \tag{4.16}$$

where $\boldsymbol{x}_T \in \mathbb{R}^{1\times L\times K_2}$ represents the compressed semantic text information, with K_2 being the dimension after the text channel encoder such that $K_2 < K_1$, and $\boldsymbol{\beta}_T$ denoting the trainable parameters.

Same as the image transmitter, the power of signal $\boldsymbol{x}_T$ needs to be normalized by $\boldsymbol{x}_T/\mathbb{E}(\|\boldsymbol{x}_T\|_2)$ for wireless transmission.

4.2.1.3 Receiver

The receiver architecture is depicted in Figure 4.12. Initially, the received symbols undergo signal detection. Subsequently, the semantic information is recovered through the respective channel decoders. Finally, the text and image semantic information are integrated to derive the answers.

The received signal at the receiver, represented by $\boldsymbol{Y}$, is modeled as

$$\boldsymbol{Y} = \boldsymbol{H}\boldsymbol{X} + \boldsymbol{N}, \tag{4.17}$$

where $\boldsymbol{H}$ corresponds to the channel matrix linking the base station (BS) and the users. Here, $\boldsymbol{X} = [\boldsymbol{x}_I, \boldsymbol{x}_T]$ combines the transmitted symbols from users transmitting text and image data, while $\boldsymbol{N}$ denotes circularly symmetric Gaussian noise with elements following a zero-mean distribution and variance σ_n^2.

In this scenario, signal detection is integrated to accelerate model training and refine decision-making precision. By employing an LS signal detector, the transmitted signal can be estimated as

$$\begin{aligned} \widehat{\boldsymbol{X}} &= \left(\boldsymbol{H}^H\boldsymbol{H}\right)^{-1}\boldsymbol{H}^H\boldsymbol{Y} \\ &= \boldsymbol{X} + \widehat{\boldsymbol{N}}, \end{aligned} \tag{4.18}$$

where $\widehat{\boldsymbol{X}} = [\hat{\boldsymbol{x}}_I, \hat{\boldsymbol{x}}_T]$ contains the estimated data for text and image users, and $\widehat{\boldsymbol{N}} = \left(\boldsymbol{H}^H\boldsymbol{H}\right)^{-1}\boldsymbol{H}^H\boldsymbol{N}$. The operation in (4.18) converts the multiplicative interference into an additive form, simplifying the learning process by significantly lowering its computational complexity.

Following signal detection, the estimated signals undergo semantic recovery via the respective channel decoders for text and image, denoted as

$$\hat{\boldsymbol{z}}_I = C_I^{-1}\left(\hat{\boldsymbol{x}}_I; \gamma_I\right), \tag{4.19}$$

and

$$\hat{\boldsymbol{z}}_T = C_T^{-1}\left(\hat{\boldsymbol{x}}_T; \gamma_T\right), \tag{4.20}$$

where γ_I and γ_T represent the corresponding trainable parameters of the image and text channel encoder, respectively.

The semantic decoder integrates the text and image semantic information to execute the visual question-answering task. This process can be mathematically represented as

$$\boldsymbol{a} = \mathcal{S}^{-1}\left(\left(\hat{\boldsymbol{z}}_I, \hat{\boldsymbol{z}}_T\right); \varphi\right), \tag{4.21}$$

where $\boldsymbol{a}$ is the predicted answer, and $\mathcal{S}(\cdot; \varphi)$ denotes the semantic decoder with trainable parameters φ.

4.2.1.4 Model Description

Semantic Codec: The image semantic encoder utilizes a pre-trained ResNet-101 architecture, with its first 30 blocks (optimized on ImageNet) extracting semantic features. These parameters remain frozen during training to retain their robust representational power from prior exposure to large-scale datasets. The text semantic encoder employs a Bi-LSTM network. Meanwhile, the semantic decoder

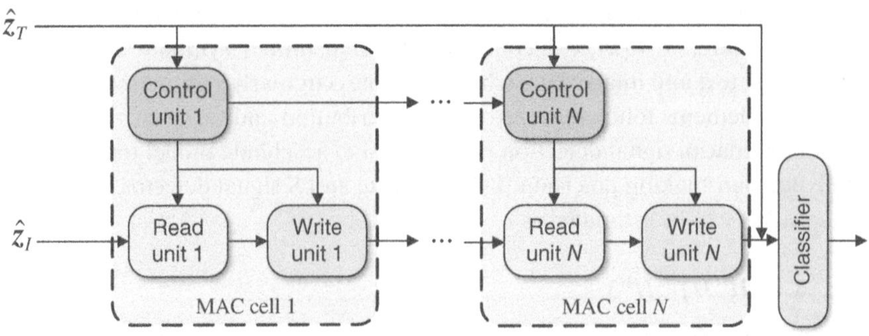

Figure 4.13 The structure of the memory, attention, and composition network for text and image semantic information fusion and the VQA.

is implemented via the MAC network [11] (Figure 4.13), which integrates multiple MAC cells. Each cell contains three components, including a control unit that generates queries from text semantics to identify question types and objects via attention, a read unit that retrieves relevant keys from image semantics using additional attention mechanisms, and a write unit that synthesizes information to produce answer predictions.

Channel Codec: The image channel encoder leverages CNN layers to capture localized image features, using convolutional layers with varying configurations for decoding. For text, the channel encoder applies stacked dense layers to retain semantic integrity, while the decoder reconstructs compressed data via dense layers.

4.2.1.5 Loss Function

As outlined earlier, the core aim of the framework is to generate precise answers to queries using input images and textual data. Unlike conventional communication systems that prioritize reconstructing complete images or texts at the receiver end, the system adopts a task-focused approach, bypassing full reconstruction to instead produce answers directly. This shift renders traditional performance metrics, such as bit-error or symbol-error rates, ineffective for evaluation. To optimize answer accuracy, the system employs the CE loss function, which quantifies the mismatch between the ground-truth answer a and the predicted output $\hat{a}$. The CE loss is mathematically expressed as

$$\mathcal{L}_{\text{CE}} = -p(a)\log\left(p(\hat{a})\right), \tag{4.22}$$

where $p(a)$ corresponds to the true distribution of the answer, and $p(\hat{a})$ represents the predicted probability distribution. This loss function effectively captures the

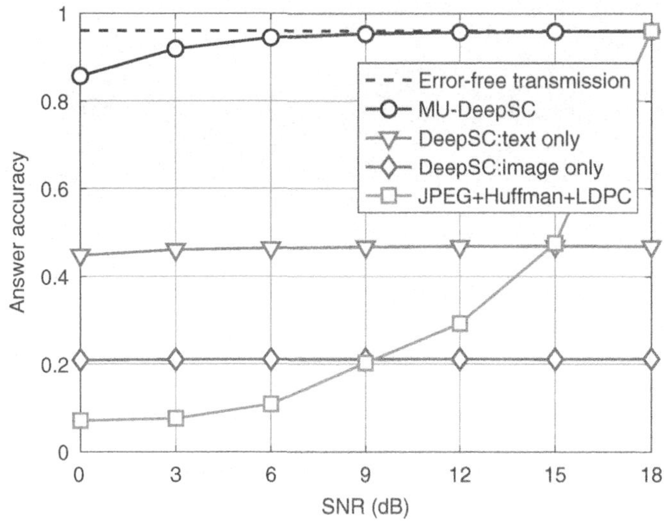

Figure 4.14 Answer accuracy comparisons between different methods over AWGN channels.

divergence between the two distributions. By minimizing the CE loss during training, the model learns to align its predictions with the correct answers, thereby maximizing the likelihood of accurate results.

4.2.1.6 Results

Figure 4.14 presents a comparative analysis of answer accuracy versus SNR over AWGN channels. As illustrated, the system consistently surpasses other baseline methods, especially in low-SNR scenarios, while closely approximating the upper-bound accuracy at higher SNR levels. In contrast, conventional methods exhibit degraded performance at lower SNR levels, largely due to image distortion caused by bit-errors; however, their accuracy improves as SNR increases. Notably, transmitting a single data modality (e.g., text or image alone) yields certain accuracy across all SNR ranges. This is because the network gives the answer from limited sets.

A subset of these outcomes is visualized in Figure 4.15. The system correctly answers all queries, whereas traditional communication systems and text-only DeepSC variants provide accurate responses for only a subset of questions. The enhanced efficacy of text-only transmission stems from the ability of text-based data to constrain the scope of potential answers, thereby improving system reliability.

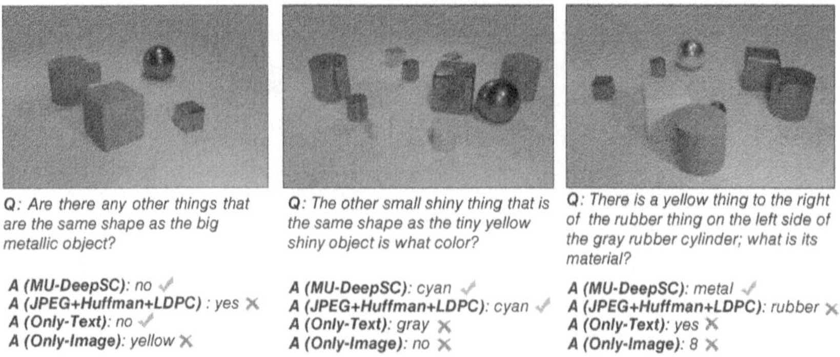

Figure 4.15 Illustrative results for the VQA task over Rician fading channels with an SNR of 18 dB. The first row displays the transmitted images, the second row presents the transmitted questions, and the subsequent four rows show the predicted answers by MU-DeepSC, traditional methods, MU-DeepSC with text only, and MU-DeepSC with image only, respectively. *Source*: Stanford University. https://cs.stanford.edu/people/jcjohns/clevr/ last accessed Sep 08, 2025.

References

1 Z. Weng, Z. Qin, X. Tao, et al., "Deep learning enabled semantic communications with speech recognition and synthesis," *IEEE Trans. Wireless Commun.*, vol. 22, no. 9, pp. 6227–6240, 2023.

2 M. Schuster and K. Paliwal, "Bidirectional recurrent neural networks," *IEEE Trans. Signal Process.*, vol. 45, no. 11, pp. 2673–2681, 1997.

3 A. Graves, S. Fernández, F. Gomez, and J. Schmidhuber, "Connectionist temporal classification: labelling unsegmented sequence data with recurrent neural networks," in *Proc. Int. Conf. Mach. Learning (ICML)*, Pittsburgh, PA, USA, 2006, pp. 369–376.

4 M. Bińkowski, J. Donahue, S. Dieleman, et al., "High fidelity speech synthesis with adversarial networks," in *Proc. Int. Conf. Learn. Represent. (ICLR)*, Formerly Addis Ababa,Ethiopia, 2020.

5 B. Bessette, R. Salami, R. Lefebvre, et al., "The adaptive multirate wideband speech codec (AMR-WB)," *IEEE Trans. Speech, Audio Process.*, vol. 10, no. 8, pp. 620–636, 2002.

6 A. Balatsoukas-Stimming, M. B. Parizi, and A. Burg, "LLR-based successive cancellation list decoding of polar codes," *IEEE Trans. Signal Process.*, vol. 63, no. 19, pp. 5165–5179, 2015.

7 A. Baevski, Y. Zhou, A. Mohamed, and M. Auli, "Wav2vec 2.0: a framework for self-supervised learning of speech representations," in *Proc. Adv. Neural Inf. Proces. Syst. (NeurIPS)*, online, 2020, pp. 12449–12460.

8 "IEEE standard for floating-point arithmetic," *IEEE Std 754-2019 (Revision of IEEE 754-2008)*, pp. 1–84, 2019.

9 H. Xie, Z. Qin, G. Y. Li, and B.-H. Juang, "Deep learning enabled semantic communication systems," *IEEE Trans. Signal Process.*, vol. 69, pp. 2663–2675, 2021.

10 J. Shen, R. Pang, R. J. Weiss, et al.,, "Natural TTS synthesis by conditioning wavenet on mel spectrogram predictions," in *Proc. IEEE Int. Conf. Acoust., Speech, Signal Process. (ICASSP)*, Calgary, AB, Canada, 2018, pp. 4779–4783.

11 D. A. Hudson and C. D. Manning, "Compositional attention networks for machine reasoning," in *Proc. Int. Conf. Learn. Represent. (ICLR)*, Vancouver, BC, Canada, 2018.

5

Joint Sensing and Semantic Communications

In this chapter, semantic sampling and sampling and coding are introduced for joint semantic perception. Specifically, there are two subsections in this chapter: the first subsection introduces the basic methods of semantic sensing and sampling, and illustrates the compressed perception-based method as an example, and the second subsection introduces semantic perception as well as joint sampling-semantic-channel coding and decoding in detail, and introduces practical applications in immersive communication and holographic video, in addition to providing the relevant experimental results under the influence of different factors, in order to demonstrate the superiority of semantic sensing and sampling in reducing transmission cost and ensuring performance under large-scale data.

5.1 Introduction and Framework of Joint Sampling and Coding

Compressed sensing is one of the common methods for downsampling, which focuses on mapping the signal to the sparse domain and then reconstructing the complete signal with fewer measurements. Traditional signal processing follows the Nyquist sampling theorem, which requires a sampling rate of at least twice the highest frequency of the signal. Compressed sensing, on the other hand, reconstructs all the information required to reconstruct the signal at a much smaller sampling rate through a sampling matrix. With the development of deep learning techniques, compressed perception combined with artificial intelligence senses and samples raw bulk-level data based on the semantic information of images, video, audio, or three-bit representations. Semantic-driven sampling and reconstruction techniques can effectively improve performance while reducing the amount of data.

Deep Learning Enabled Semantic Communications, First Edition. Zhijin Qin et al.

Published 2026 by John Wiley & Sons, Inc.

Conventional image or video sampling systems based on the Nyquist–Shannon sampling theorem require the source sampling ratio to be no less than twice the bandwidth of the original signal. However, the holographic acquisition equipment is expensive, with high sensing costs per acquisition, and keeping full sampling all the time will reduce the equipment's lifetime. In addition, targeting the transmitted information with appropriate downsampling selection is also beneficial in improving the accuracy of subsequent coded feature extraction and the performance of the transmission. To reduce the sensing overhead and improve the performance and security of semantic communication, image compression sensing is designed with the joint semantic realization of the sampling and downsampling processes as a new paradigm for image acquisition, transmission, and reconstruction. In order to maintain the end-to-end integrity of the system, a matching reconstruction algorithm is designed on the receiver side to realize the integrated design of sampling and reconstruction.

5.1.1 Semantic Sampling

In wireless multimedia transmission system based on semantic communication, collecting information with a high semantic importance level according to a specific strategy and ignoring those parts without enough semantic information can effectively reduce the sensor sensing overhead and improve the transmission efficiency, so that the information with high semantic fidelity can be more accurately extracted and encoded by the subsequent joint source-channel coding (JSCC) encoder. The adaptive channel environment is conducive to further joint source and channel characteristics. Therefore, the focus needs to be on the sampling part.

In the field of semantic sampling, a content-aware sensing approach was introduced within the framework of compressed sensing [1], enabling the fine-tuning of sensing operations through a combination of uniform and semantic matrices. Song et al. proposed a spatially adaptive sampling method that leverages prior quality maps and spatial importance to enhance spatial dependencies in quantized latent representations. This approach is further refined through probabilistic modeling, which optimizes the rate-task loss function to preserve the image quality of critical regions [2]. Building upon this foundation, Zhang et al. enhanced the attentional mechanism to integrate spatially intelligent perception [3].

Another significant research direction in downsampling pertains to computational imaging, which focuses on the intelligent design of camera-coded apertures for image and video content acquisition. Bacca [4] proposed an adjustable real-valued coded aperture in a depth optical imaging system, tailored to semantic information and specific task objectives, incorporating a regularization penalty term to optimize performance. Similarly, Vargas et al. [5] and Marwah et al. [6] conceptualized captured images as angular projections of the incident light field and designed a composite coding framework integrating a time-varying coded aperture with a spatially coded shutter. This design significantly enhanced imaging quality while ensuring temporal and spatial pixel stability.

As for semantic sampling, the most common method of semantic-aware sampling is compressed sensing. Each frame of the video, or each image, $\boldsymbol{s}_i$ is segmented into non-overlapping blocks, and each block corresponds to a specific sampling matrix $\boldsymbol{A}_{q_i}$ when sampling is performed. After the overall sampling rate of the source media is determined, different sampling rates are assigned to each block based on the SNR and the semantic information of the source through the semantic scanning network with the correction algorithm, in which the semantic scanning network consists of neural network layers. The semantic scanning network completes the channel environment adaptive semantic information importance assessment for each region of the frame, then normalizes and corrects it to obtain the sampling matrix for each block, and the semantic sampling ratio allocation map $\boldsymbol{R}$, which is shared with the receiver of the system. $\boldsymbol{R}$ shows the specific numbers of samples allocated for each block. The procedure of deep learning compressed sensing-based semantic sampling can be described as

$$\boldsymbol{w}_i = \boldsymbol{A}_{q_i}\boldsymbol{s}_i, \tag{5.1}$$

where $\boldsymbol{A}_{q_i}$ is the sensing matrix for one block, and $\boldsymbol{w}_i$ is the initial sampled result.

In order to align the sampling rate of each block with the transmission module, the initial sampled result should be processed by a generalized inverse matrix of the original sampling matrix, which is defined as $\boldsymbol{A}_{q_i}^T$, thus, it can derive

$$\boldsymbol{z}_i = \boldsymbol{A}_{q_i}^T\boldsymbol{w}_i, \tag{5.2}$$

where $\boldsymbol{z}_i$ is the symbol to be transmitted.

5.1.2 Semantic Reconstruction

Corresponding to the sampling at the sending end, the reconstruction at the receiving end adopts the deep reconstruction algorithms method based on the semantic information as well, completing the integrated matching. Its inputs are the outputs of the decoder and the semantic sampling map $\boldsymbol{R}$ shared with the sending end. The high-fidelity recovery can be accomplished after selecting a certain number of iteration rounds. In the meantime, the cost of restoration and transmission of the sampling map $\boldsymbol{R}$ between transmitter and receiver should be considered.

5.2 Joint Semantic Sampling and Coding for Image

In this section, semantic sampling and the related work of joint sampling-semantic-channel coding for images is introduced specifically. Finally, the transmission scenario of holographic video in immersive communication is combined with the application of this technique.

5.2.1 Semantic-aware Image Compressed Sensing

In automatic driving, industrial internet and holographic communication networks, the timeliness of communication (i.e., considering the optimization of communication latency) is extremely critical, and thus lightweight transmission of data is required, and downsampling the source data according to the formulation of different strategies and schemes is conducive to reducing the overhead burden of transmission and processing. Common sampling ideas to reduce overhead include sampling based on age of information (AoI), value of information (VoI), relevance, information priority or importance, as well as semantic distillation, sparse sampling, and other means to reduce the transmission content while preserving semantics. Meanwhile, recent research on semantic communication has shown that data transmission efficiency can be improved if communication strategies are modified based on the semantic information in the data. Thus, the data acquisition process is improved in a semantically aware way.

For semantic-based sampling and reconstruction techniques, there are two schemes, fixed downsampling rate and non-fixed downsampling rate, which are more generalizable in real-world contexts and are therefore introduced. Semantic sampling can be viewed as the process of selectively obtaining lightweight sources based on semantics. The encoder first uses a fixed number of elementary measurements to perceive different images (i.e., elementary uniform sampling). Based on this result, in turn, a policy network is used to analyze the semantic information in the images and determine the measurement matrices for different regions of the images (i.e., semantic sampling). At the decoder side, a semantic-aware initial reconstruction network is utilized to handle the changes in the measurement matrix at the encoder side and then downsample according to the different measurements. At the same time, rate-distortion training loss is further introduced to dynamically adjust the average compression ratio of the semantic-aware system, and utilize a trade-off loss function to jointly train the policy network with the encoder and decoder end-to-end. Since the space of the semantic-aware sampling network is much smaller than that of the image signal space on the measurement space, it has the advantage of being lightweight in scenarios such as holographic communication, where the data volume is large and the network burden is heavy.

In [1], B. Zhang et al. presents a novel semantic-aware image compressive sensing system, where the best measurement matrices for different images are decided by the images' semantic information.

5.2.1.1 Block-based Uniform Sampling

The basic process of semantic sampling is shown in Figure 5.1, where the first step is to sample the image as a whole uniformly. The visual content to be sampled (not captured by the sensor) is $I \in \mathbb{R}^{H \times W \times 3}$. Firstly, divide the image into

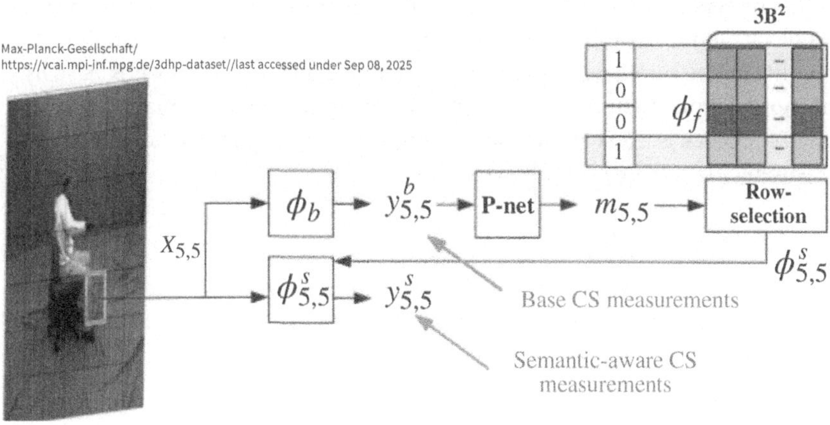

Figure 5.1 The process of block-based image sampling.

non-overlapping blocks of size and reshape the blocks into vectors of length $N = B \times B \times 3$, and each block is uniformly sampled using a learnable basic measurement matrix ϕ_b with size of $n_b \times 3B^2$, and the whole process can be described as the following equation:

$$y^b_{i,j} = \phi_b x_{i,j}, \tag{5.3}$$

where $i = 1, 2, \ldots, \frac{H}{B}$, then the number of blocks into which the visual content I can be divided is $L = \frac{H \times W}{D \times D}$; for a certain given block of I, the number of samples is n, then the compression rate of that block is counted as $\frac{n}{3B^2}$. The measurement value of block (i, j)-th under the basic measurement matrix ϕ_b is $y^b_{i,j}$.

5.2.1.2 Semantic-aware Sampling and Reconstruction

After obtaining the basic measurement value by uniform sampling, the second step is to selectively sample the image based on semantics; the complete shared measurement matrix is $\phi_f \in \mathbb{R}^{n_{\max} \times 3B^2}$, for each block of the image, according to the semantics and its relative importance to cut the matrix, obtaining the semantic measurement matrix $\phi^s_{i,j} \in \mathbb{R}^{n^s_{i,j} \times 3B^2}$ of each block. The specific process is as follows: the input is the basic measurements $y^b_{i,j}$, and the 0–1 binary mask vector $m_{i,j}$ is generated for the basic measurements of each block through the policy network $P(\cdot)$, and the process can be expressed as follows:

$$m_{1,1}, m_{1,2}, \ldots, m_{\lceil \frac{H}{B} \rceil, \lceil \frac{W}{B} \rceil} = P\left(y^b_{1,1}, y^b_{1,2}, \ldots, y^b_{\lceil \frac{H}{B} \rceil, \lceil \frac{W}{B} \rceil}\right), \tag{5.4}$$

where $m_{i,j} = \{0, 1\}^{n_{\max}}$, $n_{\max}$ is the number of rows of the shared complete measurement matrix, and the mask vector for each block $m_{i,j}$ guides the selection of the corresponding rows to construct the semantic measurement matrix for that block, and $n^s_{i,j}$ is the number of 1 in $m_{i,j}$. After basic uniform sampling and semantic-aware sampling, the number of samples for (i, j)-th block is $n_b + n^s_{i,j}$. The more samples there are, the richer and more important the semantic information of the block is. Next, each block is sampled using its corresponding semantic measurement matrix $\phi^s_{i,j}$ as follows:

$$y^s_{i,j} = \phi^s_{i,j} x_{i,j}. \tag{5.5}$$

For a dataset with N_b images, its average compression ratio is $r_{\text{avg}} = \frac{n_{\text{avg}}}{3B^2}$, where n_{avg} means the average number of samples for each block in the dataset. It can be calculated as

$$n_{\text{avg}} = \frac{1}{N\left\lceil \frac{HW}{B^2} \right\rceil} \sum_{k=1}^{N} \sum_{i=1}^{\left\lceil \frac{H}{B} \right\rceil} \sum_{j=1}^{\left\lceil \frac{W}{B} \right\rceil} \left(n_b + n^{s,k}_{i,j}\right). \tag{5.6}$$

Since the semantic measurement matrices for different blocks of the image in semantic sampling are also different, it is difficult to reconstruct all blocks through a unified large reconstruction matrix. In the reconstruction stage, different reconstruction matrices need to be designed for different blocks. First, a weight generation network $A(\cdot)$ is used to generate a reconstruction matrix $\Theta^s_{i,j}$ for each block. The input is the sampling value of the basic uniform sampling in the first step $y^b_{i,j}$. The process is shown in the following equation:

$$\Theta^s_{1,1}, \Theta^s_{1,2}, \ldots, \Theta_{\left\lceil \frac{H}{B} \right\rceil, \left\lceil \frac{W}{B} \right\rceil} = A\left(y^b_{1,1}, y^b_{1,2}, y^b_{\left\lceil \frac{H}{B} \right\rceil, \left\lceil \frac{W}{B} \right\rceil}\right). \tag{5.7}$$

The reconstruction process is shown in the following equation:

$$\hat{x}_{i,j} = \left[\Theta_b, \Theta^s_{i,j}\right] \begin{bmatrix} y^b_{i,j} \\ y^s_{i,j} \end{bmatrix}, \tag{5.8}$$

where $\Theta^s_{i,j}$ is the initial reconstruction matrix for the (i, j)-th block, Θ_b is the initial reconstruction matrix for the basic uniform sampling. Further, $\Theta^s_{i,j}$ is decomposed into a complete reconstruction matrix Θ_s shared by all blocks, and a small learnable reconstruction matrix $\tilde{\Theta}^s_{i,j}$, $n_{\max}$ is the maximum number of samplings in the second step, so the equation can be rewritten as follows:

$$\hat{x}_{i,j} = \left[\Theta_b, \Theta_s\right] \begin{bmatrix} I_{n_b \times n_b} & 0 \\ 0 & \tilde{\Theta}^s_{i,j} \end{bmatrix} \begin{bmatrix} y^b_{i,j} \\ y^s_{i,j} \end{bmatrix}. \tag{5.9}$$

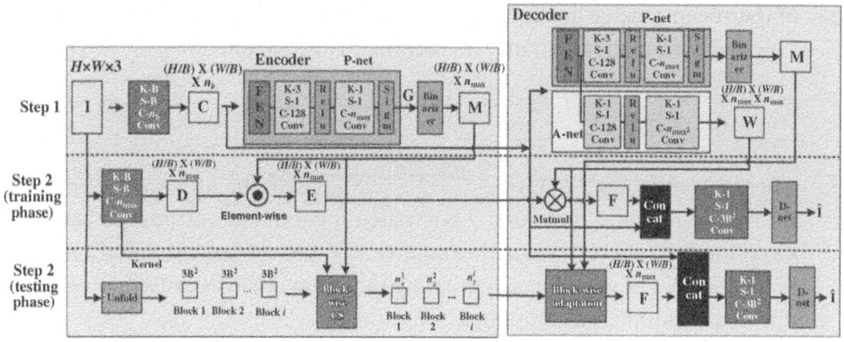

Figure 5.2 The architecture of semantic-aware sensing system.

The block-based semantic sampling and reconstruction network is shown in Figure 5.2. $K-A, S-B, C-nConv$ represents a convolution layer with a convolution kernel size of $A \times A$, a stride of B, and an output channel of n. The blue convolution layer is trained without bias parameters, and the yellow convolution layer is trained with bias parameters. The specific details are as follows:

- *Basic uniform sampling*: The image I is input into a convolution layer with a convolution kernel size of $B \times B$, a stride of B, and an output channel of n_b, and the output generated features $C \in \mathbb{R}^{\frac{H}{B} \times \frac{W}{B} \times n_b}$ are the basic sampling results.
- *Semantic sampling*: First, C is inputted into the policy network $P(\cdot)$, the intermediate features are output, and then quantized to obtain a 0–1 mask matrix. The detailed architecture is shown in Figure 5.2, where feature extraction network (FEN) represents a feature extraction network. Secondly, binary quantization is performed. If it is greater than 0.5, it is quantized to 1; otherwise, it is quantized to 0. Due to the binarization, the backward gradient is zero everywhere. Finally, multiply by the mask matrix to obtain the sampling result, where this is equivalent to semantic sampling using the selected semantic measurement matrix.
- *Semantic reconstruction*: At the reconstruction end, the basic uniform sampling value is the first input, which shares the same architecture and parameters as the encoder. Then, the output of the FEN is used as the input of the weight generation network for reconstruction, which generates weights. Matrix multiplication is performed on, and this process is repeated for all blocks, resulting in features. After obtaining these features, the uniform sampling results are connected along the channel dimension.
- *Loss function*: In the above semantic sampling and reconstruction process, the number of sampling times should be as small as possible, while the quality

of the reconstructed image should be as high as possible. Therefore, a loss function is designed to balance this conflict. The loss function is shown below:

$$L = \sum_{I \in \text{Dataset}} L_2(I, \hat{I}) + \gamma L_R(I), \tag{5.10}$$

where $L_2 = \| I - \hat{I} \|_2^2$, $L_R(I) = \sum_{i=1}^{H/B} \sum_{j=1}^{W/B} \sum_{k=1}^{n_{\max}} G(I)[i, j, k]$, $G(I)$ represents the outputs of policy network, γ is a trade-off parameter introduced between sampling loss and reconstruction quality. For the regularization term related to the number of samplings, increasing the value of γ will cause more penalties for the number of samplings.

The peak signal-to-noise ratio (PSNR) versus the average compression ratio r_{avg} of different sampling methods is shown in Figure 5.3.

As depicted in Figure 5.4, a greater number of measurements are typically allocated to image regions containing richer semantic information, such as areas depicting human subjects, illuminated regions, and green screens with folds. This adaptive allocation suggests that regions with higher semantic relevance receive increased sampling resources, thereby enhancing the preservation of critical structural and textural details during reconstruction.

5.2.2 Adaptive Sampling and Semantic-channel Coding

In [7], Z. Qi et al. presents an adaptive joint sampling-semantic-channel coding (Adaptive-JSSCC) framework, which synergistically combines semantic coding with sampling techniques. As illustrated in Figure 5.5, the framework derives region-specific optimized sampling matrices based on semantic importance, accompanied by a semantic sampling ratio distribution map, which is then shared with the receiver. This map serves as a guide for the reconstruction process, ensuring the preservation of high-quality output. Moreover, the authors introduce an attention-based Channel Adaptive Module (ACAM) to mitigate the discrepancies

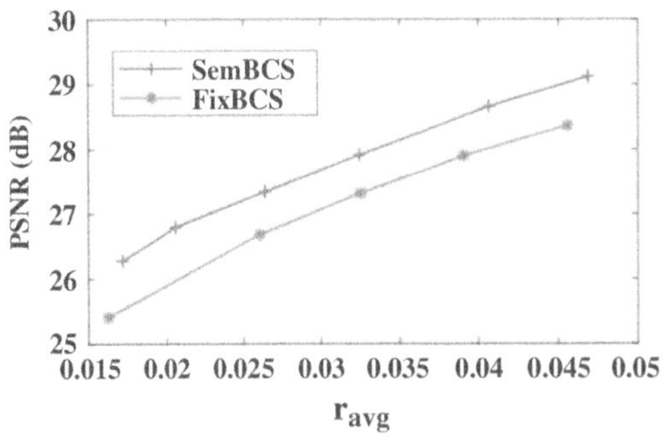

Figure 5.3 PSNR versus average compression ratio for different methods.

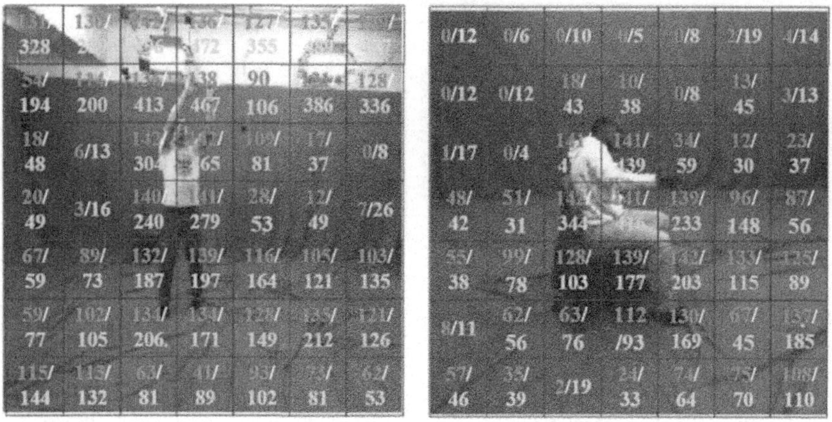

Max-Planck-Gesellschaft/https://vcai.mpi-inf.mpg.de/3dhp-dataset//last accessed under Sep 08, 2025

Figure 5.4 Examples of the learned number of measurements.

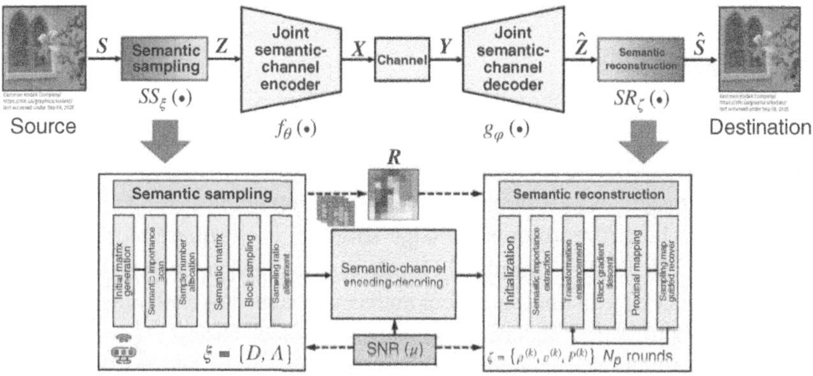

Figure 5.5 The framework of Adaptive-JSSCC, which contains semantic sampling, joint semantic-channel codex, and reconstruction.

between training and testing channel conditions in neural network models during the sampling, reconstruction, and encoding-decoding stages. To address this challenge, the signal-to-noise ratio (SNR) is incorporated as an additional input parameter, facilitating the integration and reorganization of intermediate features.

To accommodate the scaling of sampling outcomes or the representation of encoded and decoded symbols in response to fluctuating communication quality, an ACAM is designed. This module consists of several neural network layers that account for the dynamic SNRs in the prevailing environment, with its architecture depicted in Figure 5.6. The ACAM effectively captures the interplay

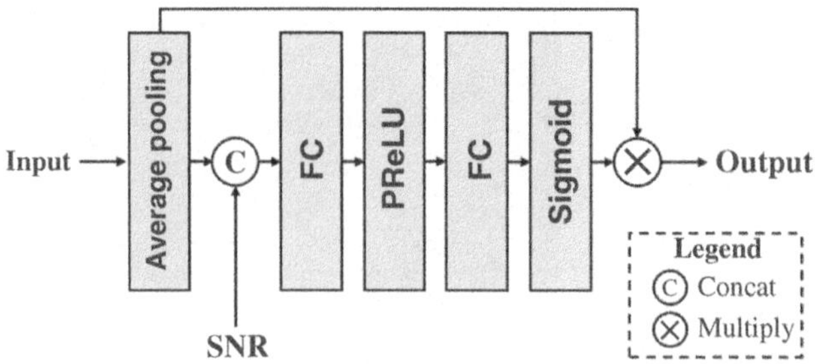

Figure 5.6 The architecture of ACAM.

between input features and channel conditions, generating distinct scaling parameters for channel characteristics through the use of channel-specific soft attention, as described in [8]. This strategy ensures the preservation of high-quality performance throughout both the sampling-reconstruction and encoding-decoding processes, even under varying SNR conditions.

5.2.2.1 Semantic Sampling

Given a predefined overall sampling ratio r, a strategy is employed to prioritize regions containing more semantically significant information, allocating a higher number of samples to these areas accordingly.

Initially, the image S is partitioned into non-overlapping blocks $\{\boldsymbol{s}_i\}_{i=1}^{l}$, each of size $N = B \times B$, where B represents the edge length of each square block. For an image of size $H \times W$, the total number of blocks is calculated as $l = \frac{H}{B} \times \frac{W}{B}$. During the sampling process, a distinct sampling matrix is assigned to each block, ensuring that the semantic sampling occurs on a block-by-block basis.

In order to evaluate the saliency of semantic information within each block and to emphasize the significance of different regions, a channel environment-adaptive semantic scanning network, denoted as $\mathcal{S}$, is introduced, as depicted in Figure 5.7. By inputting the initial image S and the SNR μ, the network generates a saliency distribution map $\boldsymbol{M} = \mathcal{S}(\boldsymbol{S}, \mu)$. Utilizing the overall sampling ratio r along with $\boldsymbol{M}$, the initial matrix $\boldsymbol{A}$ is derived, from which the semantic sampling matrix $\boldsymbol{A}_{q_i}$ is computed, and the individual sampling ratio for each block is determined as $r_i = \frac{q_i}{N}$.

To properly allocate the sampling ratio to each region in the image, the saliency map $\boldsymbol{M}$ is aggregated to form a semantic sampling ratio distribution map $\boldsymbol{R}$, which is then shared with the receiver. For this step, it is assumed that the transmission of $\boldsymbol{R}$ occurs without loss. The device then performs compressed sensing on the corresponding blocks, yielding the initial sampled results $\{\boldsymbol{y}_i\}_{i=1}^{l}$, and aligns the sampling ratios block by block for subsequent semantic encoding. Finally, appropriate symbols $\{\boldsymbol{x}_i\}_{i=1}^{l}$ are generated for transmission.

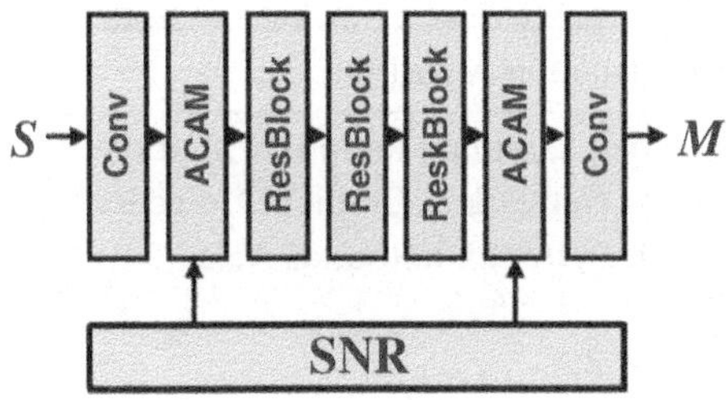

Figure 5.7 The architecture of semantic scanning network.

Algorithm 5.1: Adaptive Semantic Sampling Algorithm

Input: Overall Sampling Ratio r, Original Images $\boldsymbol{S}$, Initial Sampling Matrix $\boldsymbol{A}$ generated by SVD, SNR μ, Semantic Scanning Network $\mathcal{S}$.
Output: Semantic Sampling Ratio Distribution Map $\boldsymbol{R}$, Sampled Symbols $\boldsymbol{Z}$.
Divide $\boldsymbol{S}$ into $\{\boldsymbol{s}_i\}_{i=1}^{l}$;
$\boldsymbol{M} \leftarrow \mathcal{S}, \boldsymbol{S}, \mu$;
for *i in l s.t. overall sampling ratio r* **do**
  Obtain specific sampling ratio for each block: $r_i \leftarrow r, \boldsymbol{M}$;
  Obtain semantic sampling matrix for each block: $\boldsymbol{A}_{q_i} \leftarrow \boldsymbol{A}, r_i$;
  Sampling: $\boldsymbol{w}_i = \boldsymbol{A}_{q_i}\boldsymbol{s}_i$;
  Aligning: $\boldsymbol{z}_i = \boldsymbol{A}_{q_i}^T\boldsymbol{w}_i$;
Transform and revise: $\boldsymbol{R} \leftarrow \{r_i\}_{i=1}^{l}, \boldsymbol{M}$;
return $\boldsymbol{R}, \boldsymbol{Z} = \{\boldsymbol{z}_i\}_{i=1}^{l}$;

5.2.2.2 Semantic Reconstruction

This section provides a detailed explanation of the semantic reconstruction methodology, which is seamlessly integrated with the sampling process. It primarily involves the use of proximal gradient descent (PGD) optimization and proximal mapping, as discussed in [9]. The reconstruction process is guided by the semantic distribution map $\boldsymbol{R}$, which is transmitted to the receiver. In order to extract semantic information from $\boldsymbol{R}$, the semantic extraction network $\mathcal{E}$, structured as shown in Figure 5.7, is utilized. This network then activates the proximal mapping network to enable high-quality image reconstruction.

$$\hat{\boldsymbol{z}}_i^{(k)} = \arg\min_{\boldsymbol{z}_i} \frac{1}{2}\left\|\boldsymbol{z}_i - \boldsymbol{v}_i^{(k)}\right\|_2^2 + \lambda\mathcal{R}(\boldsymbol{z}_i), \tag{5.11}$$

where $\lambda\mathcal{R}(\boldsymbol{x}_i)$ is regularization term.

Algorithm 5.2: Adaptive Semantic Reconstruction Algorithm

Input: Decoder Output Symbols $\hat{\boldsymbol{Z}}$, Semantic Sampling Ratio Distribution Map $\boldsymbol{R}$, SNR μ, Semantic Extracting Network $\mathcal{E}$, Gradient Descent Step ρ, Proximal Mapping Network $\mathcal{P}$, Iteration Rounds N_p.

Output: Reconstructed Images $\hat{\boldsymbol{S}}$.

Initialization and enhancement: $\hat{\boldsymbol{X}}^{(0)} \leftarrow \hat{\boldsymbol{Z}}$;

for k *in* N_p **do**

  $\boldsymbol{M}' \leftarrow \mathcal{E}^{(k)}, \boldsymbol{R}, \mu$;

  Unfold into blocks: $\{\hat{\boldsymbol{z}}_i^{(k)}\}_{i=1}^{l} \leftarrow \hat{\boldsymbol{Z}}^{(k)}$;

  Gradient descent by blocks: $\{\hat{\boldsymbol{v}}_i^{(k)}\}_{i=1}^{l} \leftarrow \{\hat{\boldsymbol{z}}_i^{(k)}\}_{i=1}^{l}, \rho^{(k)}$;

  Concatenate features and reconstruct: $\{\hat{\boldsymbol{s}}_i^{(k)}\}_{i=1}^{l} \leftarrow \{\hat{\boldsymbol{v}}_i^{(k)}\}_{i=1}^{l}, \boldsymbol{M}', \mathcal{P}^{(k)}$;

  Fold and update: $\hat{\boldsymbol{Z}}^{(k+1)} \leftarrow \hat{\boldsymbol{S}}^{(k)}$;

return $\hat{\boldsymbol{S}}$;

As outlined in Algorithm 5.2, during the k-th iteration, each block undergoes an update of the proximal gradient to compute the intermediate result $\hat{\boldsymbol{v}}_i^{(k)}$. Subsequently, the sampling map $\boldsymbol{R}$, shared with the transmitter, is processed and input into the channel environment-adaptive semantic extraction network. This network generates the feature map $\boldsymbol{M}' = \mathcal{E}^{(k)}(\boldsymbol{R}, \mu)$. The intermediate result $\boldsymbol{v}_i^{(k)}$ is then concatenated with $\boldsymbol{M}'$ and passed through the proximal mapping network to yield the reconstruction result in a single iteration.

$$\hat{\boldsymbol{s}}_i^{(k)} = \hat{\boldsymbol{v}}_i^{(k)} + \mathcal{P}^{(k)}\left(\text{concat}\left(\hat{\boldsymbol{v}}_i^{(k)}, \mathcal{E}^{(k)}(\boldsymbol{R}, \mu)\right)\right), \tag{5.12}$$

where $\mathcal{P}^{(k)}$ denotes deep proximal mapping network. Due to the independent processing of image blocks, the edge effects between adjacent blocks become more pronounced, leading to noticeable artifacts during the splicing process. The operator $\mathcal{P}^{(k)}$ is designed to comprehensively capture the inter-block relationships and effectively mitigate the additional noise introduced by the splicing artifacts. The final iteration outputs $\{\hat{\boldsymbol{s}}_i^{(N_p)}\}_{i=1}^{l}$, which represent the ultimate reconstruction results.

The loss function employed for end-to-end optimization is formulated as follows:

$$\mathcal{L}_{\boldsymbol{\Pi}} = \frac{1}{lNN_b} \sum_{j=1}^{N_b} \left\| \mathcal{F}_{\text{Adaptive-JSSCC}}\left(\boldsymbol{S}_j, q_j, \mu; \boldsymbol{\Pi}\right) - \boldsymbol{S}_j \right\|_2^2, \tag{5.13}$$

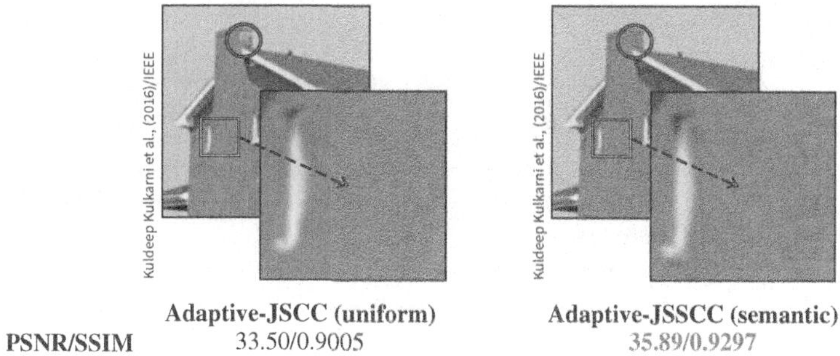

Figure 5.8 Details with uniform (left) and semantic (right) method of sampling-reconstruction.

where the set of all trainable parameters is denoted as $\Pi = \{S, A\} \cup \Theta \cup \Phi \cup \{\rho, \mathcal{E}, \mathcal{P}\}$, while $\mathcal{F}_{\text{Adaptive-JSSCC}}(\cdot)$ represents the end-to-end transmission function through the Adaptive-JSSCC framework.

5.2.2.3 Performance of Adaptive-JSSCC

As illustrated in Figure 5.8, the performance of semantic-based and semantic-free are analyzed and compared by sampling and reconstruction under a sampling ratio of 0.50 and a SNR of 3 dB. The incorporation of semantic information in the reconstruction process enhances the quality of restoration, facilitating the recovery of finer details, particularly along edges and textures. This demonstrates the advantage of semantic guidance in preserving structural integrity and visual fidelity.

Figure 5.9 presents a comparative analysis between Adaptive-JSSCC and the uniform method under varying compression levels while maintaining a constant SNR of 15 dB and an overall sampling ratio of 0.50. The channel bandwidth ratio (CBR) quantifies the residual compression space following encoding. A reduction in CBR is associated with a degraded peak signal-to-noise ratio (PSNR) and structural similarity index (SSIM) in the reconstructed image, indicating a decline in transmission quality. Furthermore, it is observed that models trained at lower CBR values exhibit earlier performance saturation, suggesting a constrained capacity for further improvement in image reconstruction quality.

5.3 Joint Semantic Sampling and Coding for Video

As illustrated in Figure 5.10, the core concept of a compressed sensing system for video involves adapting compression ratios across different spatial regions. In a conventional setup, a uniform compression ratio is applied to all spatial locations.

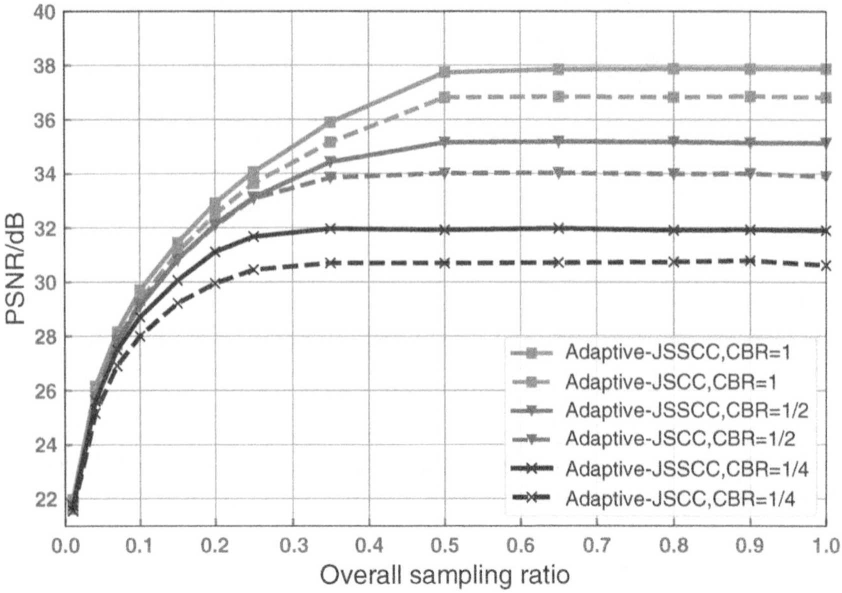

(a) PSNR performance varies in CBRs

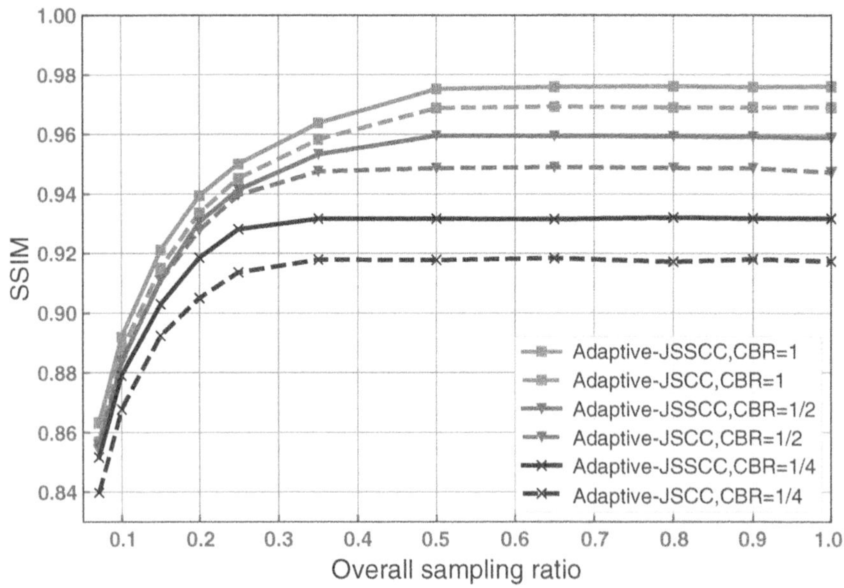

(b) SSIM performance varies in CBRs

Figure 5.9 Performance comparison between semantic and uniform sampling-reconstruction under different CBRs when the sampling ratio varies: (a) PSNR performance varies in CBRs and (b) SSIM performance varies in CBRs.

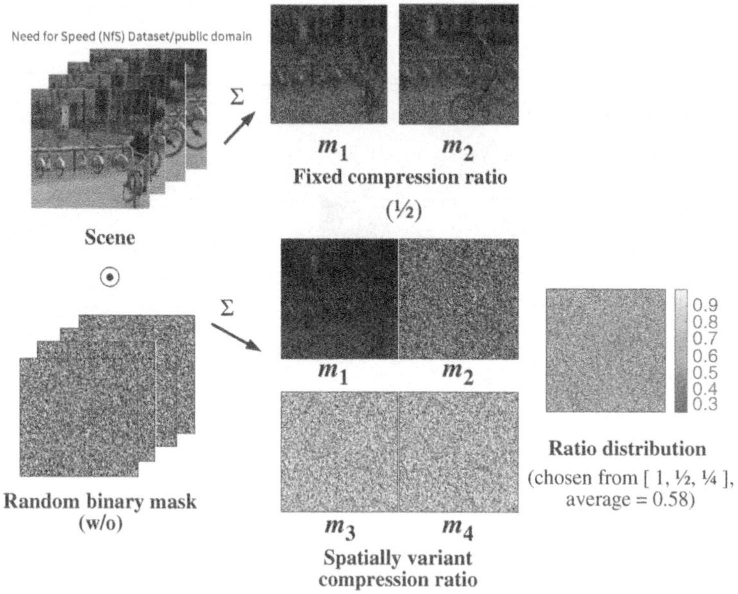

Figure 5.10 The illustration of spatially variant compression ratios in video compressed sensing.

However, in a system where the compression ratio varies spatially, certain locations receive more measurements than others. For instance, in a scenario with four video frames, a traditional snapshot compressive imaging system operating at a 1/2 compression ratio will capture two snapshots (m_1, m_2). In contrast, a system with an adaptive compression ratio would generate one output with densely sampled measurements (m_1), while the remaining three outputs (m_2, m_3, m_4) would contain more sparsely sampled data. The adaptive strategy is confirmed by semantic information in blocks or regions of the frame of the video, being captured by trained deep neural networks. More specifically, regions compressed at a 1/4 ratio will only retain measurements in m_1 after a single readout, whereas areas with a full ratio will accumulate values across all four outputs after four readout cycles. To enhance sensing efficiency, the ratio map should be co-optimized with the optical setup and the video reconstruction algorithms.

The overall framework of the video compressed sensing system is illustrated in Figure 5.11. Let H and W represent the height and width of video frames, respectively. The process begins with the generation of a compression ratio map, $\boldsymbol{M} \in \mathbb{R}^{H \times W \times 1}$, which is derived from a small trainable matrix. This matrix is expanded using a ratio generation network composed of one convolutional layer, three residual blocks, and a transposed convolution layer. Note that the spatial dimensions of the trainable matrix are set to be 1/8 of the final ratio map. Next, to

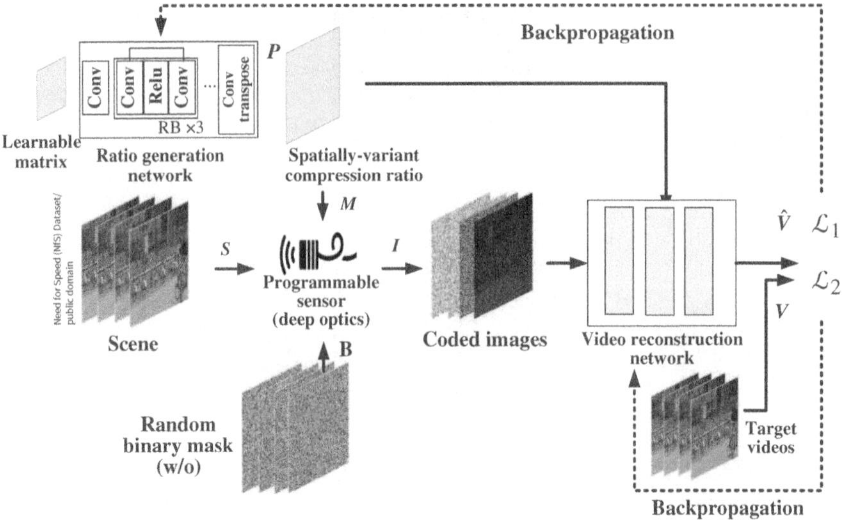

Figure 5.11 The overview architecture of the video compressed sensing system with spatially variant ratios.

capture a scene, $\boldsymbol{S} \in \mathbb{R}^{H\times W\times T}$, the process is simulated using a programmable sensor, incorporating both the $\boldsymbol{M}$ and an optional random binary mask, $\boldsymbol{B} \in \mathbb{R}^{H\times W\times T}$, that is commonly referred to as coded apertures in prior studies [5, 10]. The sensed data, I, which is the semantic-based down-sampled results of the original scenes $\boldsymbol{S}$, along with the ratio map M, is then input into a video reconstruction network to generate the final reconstructed video, $\widehat{\boldsymbol{V}} \in \mathbb{R}^{H\times W\times T}$.

To enhance reconstruction quality while simultaneously optimizing the average compression ratio, two loss functions ($\mathcal{L}_1$, $\mathcal{L}_2$) are introduced. These losses drive updates to the trainable matrix, the ratio generation network, and the video reconstruction model, ensuring an improved balance between efficiency and accuracy.

5.3.1 Semantic-based Video Sampling and Reconstruction

When the sensed data, $\boldsymbol{I}$, is acquired, the next task is to reconstruct the desired video using a deep neural network (DNN). The architecture of the video reconstruction network is depicted in Figure 5.12. This network includes three distinct modules: an initial reconstruction (IR) phase, a fusion network (FN), and a deep reconstruction network (DRN) that builds upon the single-stage spectral-wise transformers (SST) proposed in [11]. The IR and FN modules are incorporated to address the challenges posed by the spatially varying $\boldsymbol{M}$ and $\boldsymbol{B}$. Specifically, if $\boldsymbol{B}$ is excluded, the IR stage provides an initial reconstruction, $\widehat{\boldsymbol{V}}_0 \in \mathbb{R}^{H\times W\times T}$, which is denoted as

$$\widehat{\boldsymbol{V}}_{0p} = \boldsymbol{A}_{r_p}^T (\boldsymbol{A}_{r_p} \boldsymbol{A}_{r_p}^T)^{-1} (\boldsymbol{I}_p), \tag{5.14}$$

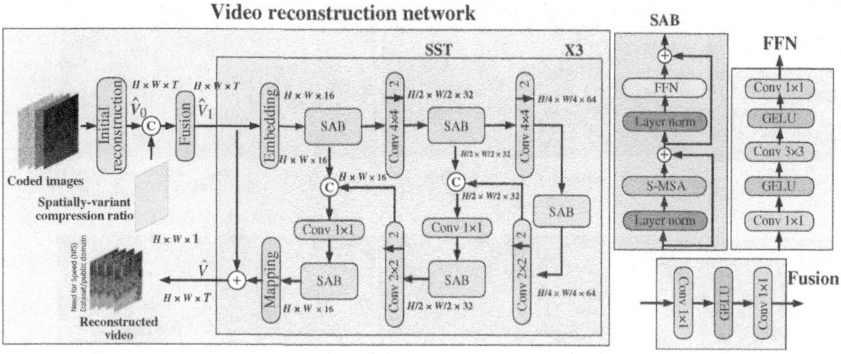

Figure 5.12 The architectural details of video reconstruction network.

where $\hat{\boldsymbol{V}}_{0p} \in \mathbb{R}^{T\times 1}$ denotes the initial reconstruction results at the spatial location p, $(\boldsymbol{A}_{r_p}\boldsymbol{A}_{r_p}^T)^{-1}$ is fortunately a diagonal matrix. When the ratio, r, is specific, the equivalent measurement matrix, $\boldsymbol{A}_r$, is of size (rT, T), with each row representing one measurement base. The i-th $(i = 1, 2, .., rT)$ row, $[\boldsymbol{A}_r]_{i:}$, is a sparse vector, defined as

$$[A_r]_{ij} = \begin{cases} 1/T, & \text{if } j \in [(i-1)/r, i/r], \\ 0, & \text{otherwise.} \end{cases} \tag{5.15}$$

In contrast, when the random binary masks, $\boldsymbol{B} \in \{0, 1\}^{H\times W\times T}$, are taken into account, the sensed data at location p, denoted as $\boldsymbol{I}_p$, can be modelled as

$$\boldsymbol{I}_p = \mathcal{U}(\hat{\boldsymbol{A}}_{r_p}\boldsymbol{S}_p), \tag{5.16}$$

where $\mathcal{U}(\cdot)$ is the camera exposure function that accounts for the noise effects introduced by the camera [12]. $\hat{\boldsymbol{A}}_{r_p} = \boldsymbol{A}_{r_p}\text{diag}(\boldsymbol{B}_p)$, where $\boldsymbol{B}$ represents a $T\times T$ diagonal matrix constructed from $\boldsymbol{B}_p$. In this scenario, some rows of $\hat{\boldsymbol{A}}_{r_p}$ may become zero, which results in $\hat{\boldsymbol{A}}_{r_p}\hat{\boldsymbol{A}}_{r_p}^T$ being non-invertible. Because the row vectors of $\hat{\boldsymbol{A}}_{r_p}$ are orthogonal to one another, $\hat{\boldsymbol{V}}_0$ can be reconstructed by independently using the non-zero row vectors of $\hat{\boldsymbol{A}}_{r_p}$.

$$\hat{\boldsymbol{V}}_{0_p} = \sum_{\substack{i=1 \\ [\hat{\boldsymbol{A}}_{r_p}]_{i:} \neq \boldsymbol{0}_{1\times T}}}^{rT} [\hat{\boldsymbol{A}}_{r_p}]_{i:}^T([\hat{\boldsymbol{A}}_{r_p}]_{i:}[\hat{\boldsymbol{A}}_{r_p}]_{i:}^T)^{-1}[\boldsymbol{I}_p]_{i:} + \boldsymbol{0}_{T\times 1}. \tag{5.17}$$

In line with the coarse-to-fine approach, a shallow FN can be applied to handle the spatially varying coded exposures and ratios. This network takes $\hat{\boldsymbol{V}}_{0_p}$, $\boldsymbol{M}$,

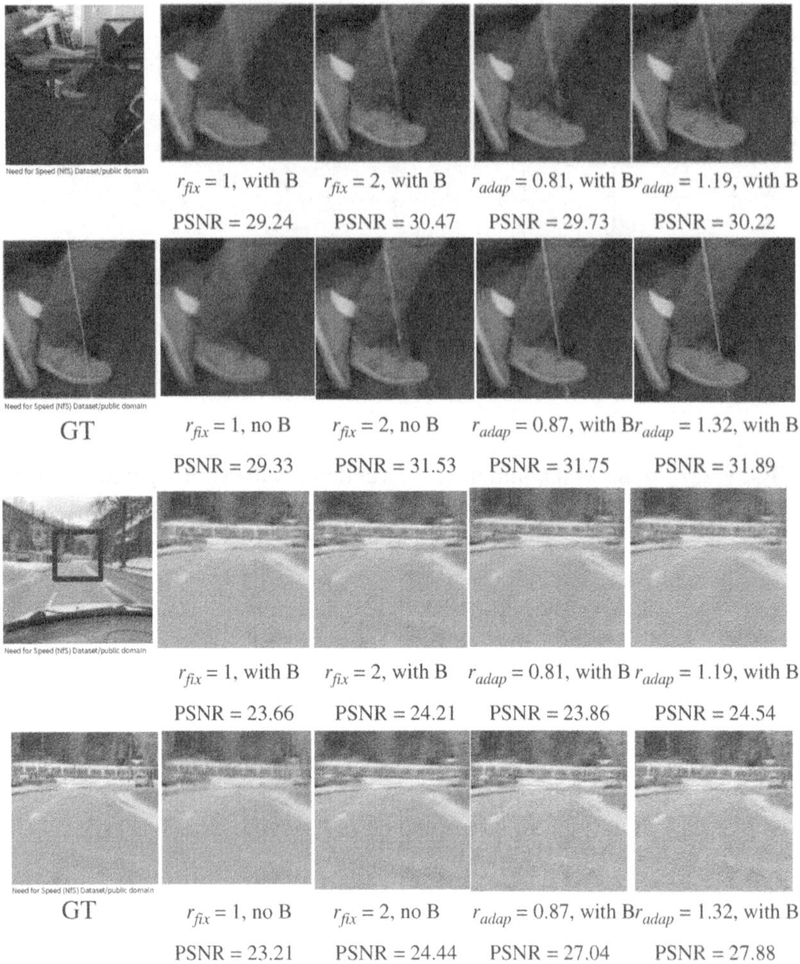

Figure 5.13 Examples of restored video frames from different sensing methods with different compression ratios.

and potentially $\boldsymbol{B}$ as inputs, producing the second-level reconstruction output, $\widehat{\boldsymbol{V}}_1 \in \mathbb{R}^{H \times W \times T}$. The structure of the FN is shown in Figure 5.12.

Subsequently, the DRN uses $\widehat{\boldsymbol{V}}_1$ as input to generate the final video, $\widehat{\boldsymbol{V}}$, by stacking three SSTs [11]. The embedding and mapping blocks consist of convolutional layers with 3×3 kernels. In the encoder, the feature maps are passed through a sequence of one spectral-wise attention block (SAB) [11], a convolutional layer with a stride of two for downsampling, another SAB, and another convolutional layer with a stride of two. The decoder mirrors the architecture of the encoder.

Skip connections are employed to aggregate features between the encoder and decoder, mitigating information loss during downsampling. The core component of the SST is the SAB. The structure of SAB is also shown in Figure 5.12. It includes a feed-forward network (FFN), a spectral-wise multi-head self-attention (S-MSA) mechanism, and two layer-normalization operations. Unlike the original MSA, which operates along the spatial dimension to compute self-attention, S-MSA treats each feature map as a token and computes self-attention across the channel dimension, which enhances computational efficiency. Further details on S-MSA can be found in [11].

The training process of the video reconstruction network is supervised, with the objective of reducing the MSE between $\boldsymbol{V}$ and $\hat{\boldsymbol{V}}$,

$$\mathcal{L}_{\mathrm{MSE}}(\boldsymbol{V}, \hat{\boldsymbol{V}}) = \sum_{p=1}^{P} (V_p - \hat{V}_p)^2. \tag{5.18}$$

5.3.1.1 Numerical Results

Figure 5.13 shows the restored video frames from different sensing methods with different average compression ratios. Each example has two rows. The first rows are the results when $\boldsymbol{B}$ is used, while the second is when $\boldsymbol{B}$ is not used. $r_{fix} = 1$ or $r_{fix} = 2$ denote the fixed-ratio method with $\bar{r} = 1$ and $\bar{r} = 2$, respectively. $r_{adap} = 0.81, 1.19$, or 1.32 denotes learned-ratio method with $\bar{r} = 0.81, 1.19$, or 1.32, respectively. From the figure, the frames from the learned ratio method have more texture details and fewer artifacts.

References

1 B. Zhang, Z. Qin, and G. Y. Li, "Semantic-aware image compressed sensing," in *Proc. IEEE Int. Workshop Mach. Learn. Signal Process. (MLSP)*, IEEE, 2023, pp. 1–6.

2 M. Song, J. Choi, and B. Han, "Variable-rate deep image compression through spatially-adaptive feature transform," in *Proc. IEEE/CVF Int. Conf. Comput. Vis. (ICCV)*, 2021, pp. 2380–2389.

3 B. Zhang, Z. Qin, and G. Y. Li, "Compression ratio learning and semantic communications for video imaging," *IEEE J. Sel. Top. Signal Process.*, vol. 18, no. 3, pp. 312–324, 2024.

4 J. Bacca, T. Gelvez-Barrera, and H. Arguello, "Deep coded aperture design: An end-to-end approach for computational imaging tasks," *IEEE Trans. Comput. Imaging*, vol. 7, pp. 1148–1160, 2021.

5 E. Vargas, J. N. Martel, G. Wetzstein, and H. Arguello, "Time-multiplexed coded aperture imaging: Learned coded aperture and pixel exposures for compressive imaging systems," in *Proc. IEEE/CVF Int. Conf. Comput. Vis. (ICCV)*, 2021, pp. 2692–2702.

6 K. Marwah, G. Wetzstein, Y. Bando, and R. Raskar, "Compressive light field photography using overcomplete dictionaries and optimized projections," *ACM Transactions on Graphics (TOG)*, vol. 32, no. 4, pp. 1–12, 2013.

7 Z. Qi, Y. Feng, and Z. Qin, "Adaptive sampling and joint semantic-channel coding under dynamic channel environment," *arXiv preprint arXiv:2502.07236,* 2025.

8 J. Xu, B. Ai, W. Chen, A. Yang et al., "Wireless image transmission using deep source channel coding with attention modules," *IEEE Trans. Circuits Syst. Video Technol.*, vol. 32, no. 4, pp. 2315–2328, 2021.

9 B. Chen and J. Zhang, "Content-aware scalable deep compressed sensing," *IEEE Trans. Image Process.*, vol. 31, pp. 5412–5426, 2022.

10 X. Yuan, Y. Liu, J. Suo, and Q. Dai, "Plug-and-play algorithms for large-scale snapshot compressive imaging," in *Proc. IEEE Comput. Soc. Conf. Comput. Vis. Pattern Recognit. (CVPR)*, Seattle, WA, USA, 2020, pp. 1447–1457.

11 Y. Cai, J. Lin, Z. Lin, H. Wang et al., "MST++: Multi-stage spectral-wise transformer for efficient spectral reconstruction," in *Proc. IEEE Comput. Soc. Conf. Comput. Vis. Pattern Recognit. (CVPR)*, New Orleans, LA, USA, 2022, pp. 19–20.

12 C. M. Nguyen, J. N. Martel, and G. Wetzstein, "Learning spatially varying pixel exposures for motion deblurring," in *Proc. IEEE Int. Conf. Comput. Photogr. (ICCP)*, Pasadena, CA, USA, 2022, pp. 1–11.

6

Semantic Impairments in Communications

Semantic communications are not only influenced by physical channel impairments during transmission but also susceptible to semantic impairments. Initially, Bao et al. [1] discussed the concept of semantic noise and characterized it as signals that trigger deviations in semantic similarity and reduce semantic infidelity. Robustness against semantic impairments is a crucial determinant of efficient semantic exchange between the transmitter and the receiver, limiting the performance of semantic information transmission. Therefore, thorough analyses of semantic impairments and specific semantic impairment suppression mechanisms tailored to diverse modal sources are necessary to enable robust semantic communications.

In particular, semantic impairments in sources take various forms, including spelling errors in text, external noise in speech, and adversarial perturbations in images. Additionally, in deep learning enabled semantic communications, the end-to-end (E2E) system is modeled by constructing multiple neural networks (NNs) and invoking appropriate loss functions to facilitate model training until stable convergence is achieved. Through this process, semantic information is conveyed and represented by different hidden layers or the output of the trained NNs. In this context, semantic impairments can be extended to the disturbances to the adopted NNs in semantic communication systems, including the extreme parameter initialization that exacerbates the system collapse, the stochastic weight perturbation that disables the training convergence, the persistent external interference that triggers the gradient vanishing or explosion, and the substantial invalid data that deviates the weight updating. However, a comprehensive theory of the NN-associated semantic impairments is lacking due to the enormous mathematical bottlenecks of deep learning (DL) techniques, and the exploration of the bespoke mechanism to suppress such semantic impairments remains elusive.

Deep Learning Enabled Semantic Communications, First Edition. Zhijin Qin et al.

Published 2026 by John Wiley & Sons, Inc.

This chapter discusses semantic impairments in communications. Specifically, the semantic impairments in different types of data are defined and categorized. Then, semantic impairment suppression mechanisms tailored for specific sources are introduced, and numerous experimental results are provided to demonstrate their effectiveness.

6.1 JSCC Framework with Semantic Impairments

Although semantic communication systems improve the efficiency of information transmission and can overcome the cliff effect issues encountered by traditional communication methods, when delving into their working principles, it is not difficult to discover that semantic communication systems also have some potential problems.

Current semantic communication systems rely on cutting-edge DL algorithms to accurately map the source data to a specific semantic space, with all subsequent operations of the communication system taking place in that space. Specifically, during the encoding phase, the system transforms the raw source data into semantic representations to delve deep and capture its intrinsic, underlying semantic information. In the decoding phase, the semantic communication system needs to interpret the semantic information and obtain the received data while ensuring that the representations of the source and received data in the semantic space are as close as possible.

However, since semantic communication systems heavily rely on DL algorithms for semantic extraction and interpretation, it is noteworthy that DL algorithms are highly sensitive to adversarial perturbations [2]. As illustrated in Figure 6.1, adding adversarial perturbations to an image of a horse may lead to the misinterpretation of the semantic information contained by DL models. Therefore, in comparison to classical communication systems, semantic communication systems not only have to address the challenges of physical impairments but also need to be cautious of the potential disruptions caused by adversarial perturbations. This heightened sensitivity to adversarial perturbations poses a greater challenge to the robustness of semantic communication systems.

To investigate the robustness of semantic communication systems in disturbed environments, academia has introduced the concept of semantic impairment [3]. Semantic impairments refer to signals that lead to semantic ambiguity between the communication transmitter and receiver and can be divided into two main categories. The first category is source semantic impairment. This type of impairment originates from a source containing incorrect semantic information, which cannot be eliminated or suppressed by traditional information symbol-based transmission methods. Figure 6.2 visually presents the manifestations of source semantic impairments in different modalities of information, including text with misspellings and images affected by adversarial perturbations. These impaired

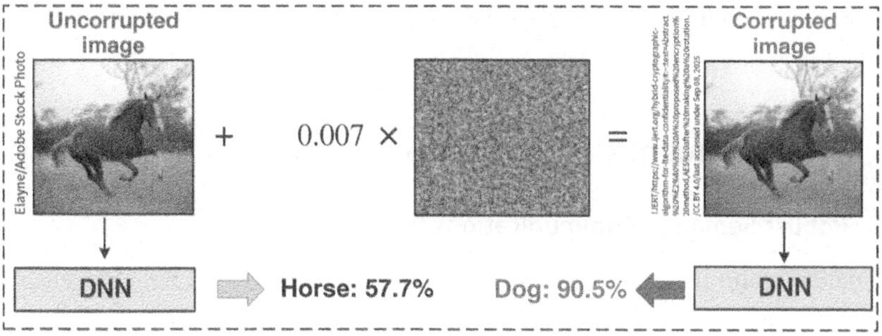

Figure 6.1 DL models are sensitive to perturbations.

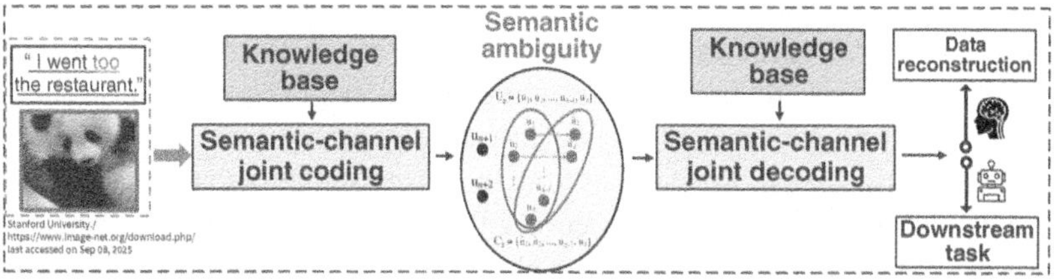

Figure 6.2 Source semantic impairments in different modalities of information.

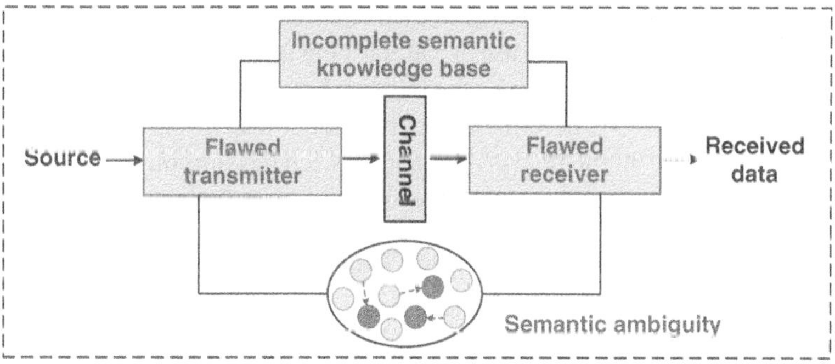

Figure 6.3 Knowledge base semantic impairments.

sources may lead to semantic ambiguity in the system, and without specific system design, they will inevitably impact semantic fidelity.

Another category is knowledge base semantic impairment, which stems from various factors, such as the incompleteness of semantic knowledge bases (SKBs), improper choices in model training methods, and unreasonable design of loss functions. These factors collectively lead to deviations in semantic representation or interpretation, thereby causing semantic ambiguity between the communication transmitter and receiver. As shown in Figure 6.3, these kinds of semantic impairments can significantly impact the accuracy and reliability of communication systems.

6.2 Source Semantic Impairments Suppression

6.2.1 Robust Semantic Communications for Text

In the field of natural language processing (NLP), research on robustness focuses on text semantic correction, which can be broadly categorized into two main types.

The first type of research is grammatical error correction (GEC), which primarily deals with correcting grammatical errors in text. Zhao et al. have devised a method based on dynamic masking techniques, which offers the advantage of achieving a more efficient correction model without the need for additional text data for training, thus enhancing the training efficiency of the model [4]. They proposed a replication-enhanced architecture for the grammar correction task, which intelligently copies correct words from the source sentence to the target sentence and enhances text correction accuracy through pre-training on unlabeled data and multi-task learning [5]. Zhang et al. cleverly utilized existing pre-trained language models for extracting text semantics and introduced an error detection network and error correction network to detect and correct semantic errors in text effectively. This method also significantly improves the accuracy and efficiency of Chinese text correction through the use of soft masking techniques [6].

The second type of research focuses on the limitations faced by systems based on deep neural networks (DNNs) in terms of generalization, which also leads to text error issues. For instance, in the field of automatic speech recognition (ASR), ASR algorithms frequently generate text errors due to factors such as recognition accuracy, background noise, and clarity of speech sources [7]. Unlike common spelling errors, ASR errors are more complex and diverse in form. For example, ASR errors may manifest as homophones, where words that sound similar but have different meanings are confused, such as the misrecognition between "sun" and "son." This complexity makes the correction of ASR errors a uniquely challenging research task.

To overcome the numerous challenges in ASR, researchers actively explore and develop various innovative algorithms to reduce the probability of recognition errors, thereby enhancing the robustness of text semantics. Li et al. improved the model's correction performance for ASR errors by using pre-trained bidirectional encoder representations from transformers (BERT) encoder and designing a decoder based on a replication mechanism [8]. Leng et al. proposed a non-autoregressive error correction model based on an edit alignment strategy, which ingeniously combines error patterns and correction operations unique to ASR, achieving faster inference speeds compared to autoregressive correction methods while maintaining high accuracy [9]. Zhang et al. developed a dual-channel ASR error correction model that detects semantic errors in the recognized source and then extracts dual-channel features from these areas. The subsequent model utilizes these features along with contextual semantic information to effectively eliminate errors in ASR [10].

Although current text semantic correction algorithms can, to some extent, eliminate erroneous semantics in text and thereby enhance text semantic robustness, these algorithms still face some significant limitations.

Firstly, noise and attenuation in the physical channel can further damage semantic information. Existing correction algorithms initially did not fully consider the impact of channel transmission on the fidelity of system semantic information, making it difficult to achieve deep integration and cooperative optimization with communication systems.

Secondly, current research on text semantic correction is generally based on the assumption that the decoder can directly access text information. This assumption allows the original text or text processed using replication mechanisms to be directly inputted into the decoder. While this simplifies the difficulty of problem-solving, this assumption does not hold in communication scenarios. The receiver in a communication scenario cannot access text information before executing the corresponding decoding operation. Therefore, existing correction algorithms often operate independently of communication systems, serving as standalone modules either at the front or back end of communication systems responsible for correcting literal errors in text to be sent or already received, as shown in Figure 6.4. This independent design not only hinders the deep integration of correction algorithms and communication systems but also leads to the unnecessary consumption of computational resources due to repeated semantic extraction and interpretation processes during text correction.

Besides, while most current text correction work adopts the classic transformer framework, there is still room for improvement in terms of feature utilization. Specifically, existing correction work typically relies on transformers for semantic feature extraction, and based on these features, subsequent text editing or generation operations are conducted to eliminate semantic errors in the text. However, there are specific issues in this process because although transformers can extract rich semantic features, these features may contain misleading information due to erroneous text. If these unfiltered and uncorrected semantic features are

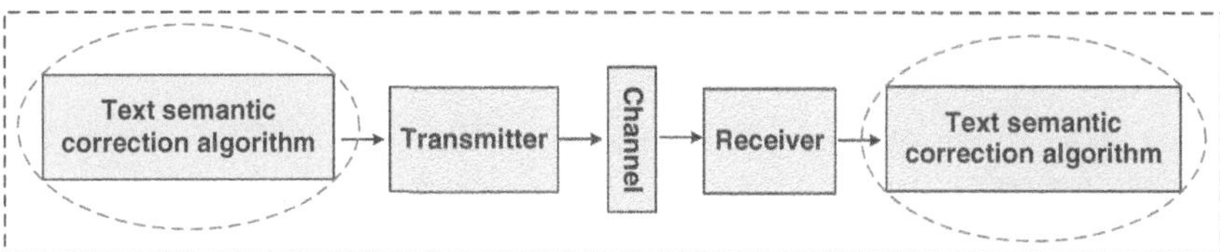

Figure 6.4 Text semantic correction methods independent of communication systems.

directly used for further processing, the correction performance of the model may be severely impacted.

Lastly, most current text correction work generally overlooks the potential impact of comprehensive types of text errors and varying degrees of semantic errors on model performance. However, in practical applications, due to the complexity of information sources and the time-varying nature of channels, the forms and degrees of text errors become exceptionally complex and diverse. Faced with this challenge, the generalization ability of DL models often appears inadequate in addressing these challenges, further making existing text correction algorithms difficult to adapt to complex and ever-changing communication scenarios.

A robust semantic communication system, referred to as R-DeepSC [3], which employs a semantic corrector for robust semantic encoding, is developed. Moreover, a speedy version, named NA-RDeepSC [3], has been proposed, which utilizes an adaptive generator and non-autoregressive architecture for significantly improving the inference speed while maintaining robustness, making it a cost-effective yet efficient solution for online services. Figure 6.5 portrays the architecture of NA-RDeepSC, which can handle both physical channel effects and semantic impairments.

Firstly, the dataset used by NA-RDeepSC will be introduced. Since literal changes in text can lead to semantic ambiguity, falling within the realm of semantic impairment, they can be categorized as human-induced spelling errors and artificial intelligence (AI) algorithm recognition errors, with AI algorithm recognition errors primarily involving automatic audio recognition errors. The datasets used are Europarl and Librispeech. Based on these corpora and some publicly available error rules and semantic communication systems, a dataset of semantic impairments is constructed.

Furthermore, the constructed dataset of semantic impairments contains various forms of errors, such as deletion, insertion, replacement, verb, boundary, and homophonic errors. A measurement of semantic impairment intensity is employed for quantifying semantic impairment. Finally, a text dataset containing different types of semantic impairments is obtained, with the semantic impairment intensity of the dataset ranging from 0 to 0.8 in intervals of 0.2.

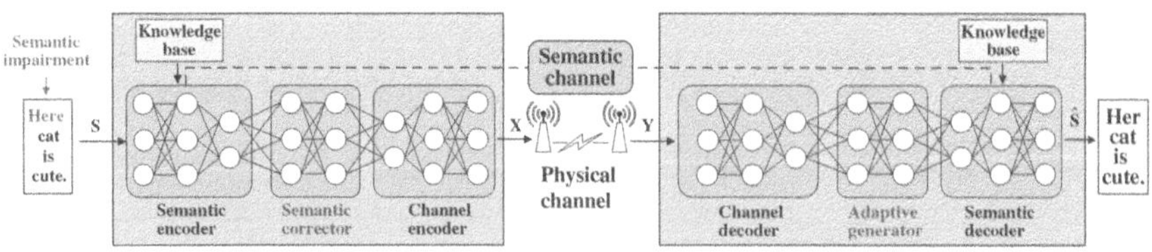

Figure 6.5 The robust semantic communication system proposed for mitigating the impact of literal semantic impairments.

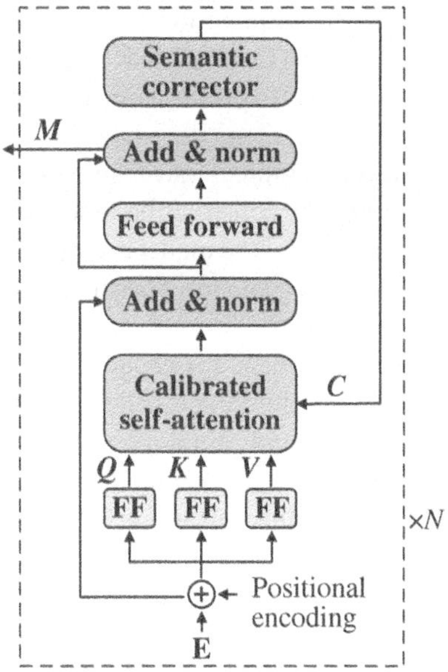

Figure 6.6 The robust semantic encoder developed, which relies on a semantic corrector and a calibrated self-attention mechanism.

Secondly, in terms of model design, NA-RDeepSC utilizes a robust semantic encoder based on a corrective self-attention mechanism, which is shown in Figure 6.6. Structurally, the semantic encoder is based on the transformer but incorporates semantic correction design and introduces feedback. Specifically, it involves three steps.

Firstly, detecting the presence of semantic impairments in the text, which can be represented as:

$$\boldsymbol{P} = f_{\epsilon}(\boldsymbol{M}), \tag{6.1}$$

where $f_{\epsilon}(\cdot)$ is the semantic corrector having the trainable parameter set ϵ, and $\boldsymbol{M}$ is the output of the semantic encoder.

Then, an attention correction matrix based on the detection results is generated, which is formulated as:

$$\boldsymbol{C} = \boldsymbol{1} - \boldsymbol{P} \cdot \boldsymbol{P}^T. \tag{6.2}$$

Finally, adjusting attention scores based on this correction matrix, which can be expressed as:

$$\boldsymbol{A}_c = \text{softmax}\left(\frac{\boldsymbol{Q} \cdot \boldsymbol{K}^T}{\sqrt{d_k}} \odot \boldsymbol{C}\right), \tag{6.3}$$

where $\odot$ represents the element-wise product, $\boldsymbol{Q}$, $\boldsymbol{K}$, $\boldsymbol{V}$, and d_k are the query, key, value, and the dimension of the encoded semantics.

This design allows the model to dynamically adjust its self-attention mechanism based on the probability of semantic impairments, ensuring that the model focuses most of its attention on text without semantic impairments to ensure the accuracy of semantic information.

Additionally, to enhance decoding speed, a non-autoregressive semantic decoder is built. Compared to the original transformer, by constructing independent conditional sequences, the decoding time complexity of the model is reduced from $O(n)$ to $O(1)$, enabling parallel decoding and reducing latency. The structure of this encoder is shown in Figure 6.7.

Two semantic communication systems are trained, where R-DeepSC uses a robust semantic encoder and autoregressive decoding, exhibiting excellent error correction performance. NA-RDeepSC, on the other hand, employs a robust semantic encoder and non-autoregressive semantic decoder, achieving fast inference speed. By effectively combining these two models, different needs for error correction performance in various scenarios can be met.

Finally, the experimental results are as follows. The experiments are conducted under different signal-to-noise ratios (SNRs), with the channels being additive white Gaussian noise (AWGN) and Rician fading channels. Under the AWGN channel, the NA-RDeepSC outperforms traditional communication systems by at least 8.1% in terms of bilingual evaluation understudy (BLEU) scores, which is

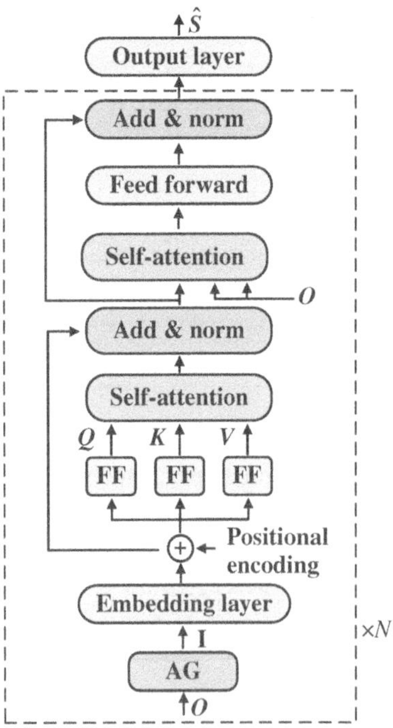

Figure 6.7 The semantic decoder developed in NA-RDeepSC.

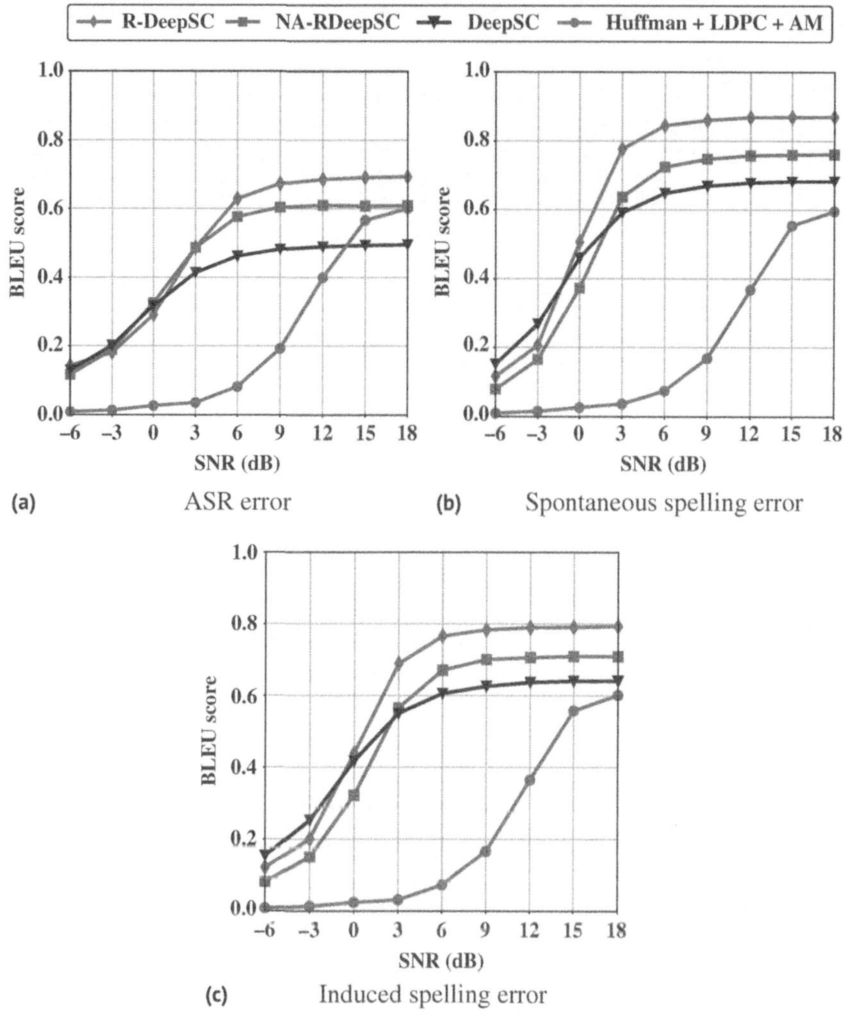

(a) ASR error (b) Spontaneous spelling error (c) Induced spelling error

Figure 6.8 System performance in AWGN channels versus the SNR with various types of semantic impairments: (a) ASR error; (b) spontaneous spelling error; and (c) induced spelling error.

shown in Figure 6.8. Specifically, at an SNR of 3 dB, the NA-RDeepSC outperforms the existing semantic communication system DeepSC.[11] by 13.6, 16.7, and 13.1% under three forms of semantic impairments.

Additionally, the experiments are also conducted under different semantic impairment intensities as shown in Figure 6.9. The NA-RDeepSC outperforms existing semantic communication systems by at least 8.1% in terms of BLEU scores and 17.3% in terms of BERT scores.

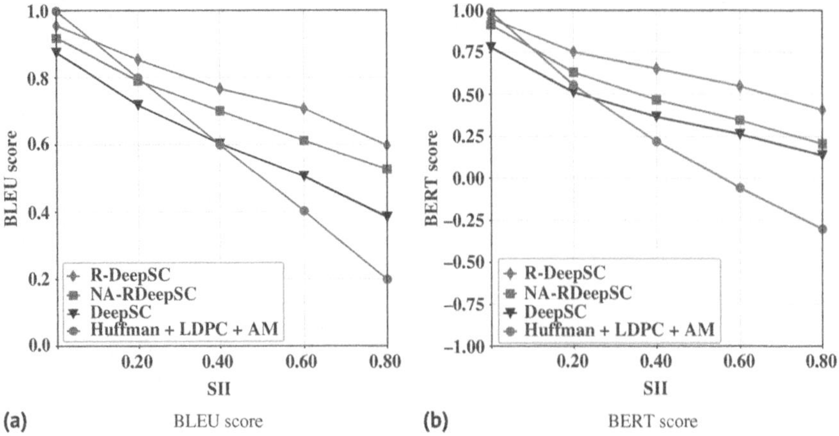

Figure 6.9 System performance versus semantic impairment intensity under AWGN channels: (a) BLEU score and (b) BERT score.

6.2.2 Robust Semantic Communications for Speech

The investigation of semantic communications in speech transmission faces challenges of semantic impairments due to the corruption of input speech. In contrast to traditional noise that affects the accuracy of received symbols, semantic impairments in speech refer to various interferences that are capable of inducing semantic ambiguity and reducing semantic fidelity. Semantic impairments in speech are pervasive in real-world communications and significantly impact the fidelity of semantic exchange, while this issue is ignored in existing research. Therefore, it is significant to consider semantic impairments in semantic communications for speech transmission.

To address this problem, a feasible solution is to remove external interference from the corrupted input speech before feeding it into the communication system. For example, when the input speech is corrupted, a speech enhancement (SE) module can be utilized before the semantic encoder to reduce semantic impairments. However, this separate architecture requires additional computational resources to reconstruct the enhanced speech and runs counter to the ultimate goal of effective E2E semantic transmission. An effective mechanism is required that estimates the damaged semantic information inherent in the corrupted speech and generates deep semantic features simultaneously. This section introduces the generative adversarial network (GAN)-enabled robust semantic communications, named Ross-S2T [12], for serving speech-to-text translation (S2TT) tasks when the input speech is corrupted.

6.2.2.1 Speech Semantic Impairments

Due to the unique nature of speech, it encompasses distinct acoustic characteristics and emotional expressions. Additionally, textual information forms a critical component of speech, requiring accurate transmission to the recipient. In real-world communication scenarios, speech is exposed to surrounding environments and is vulnerable to external interferences, such as background noise and overlapping speech, resulting in the degradation of speech input in wireless communication systems and compromising its intrinsic semantic content. Semantic impairments can be classified into acoustic semantic impairments and textual semantic impairments. An example of acoustic and textual semantic impairments is shown in Figure 6.10. From the figure, acoustic semantic impairments caused by train noise completely obscure portions of human speech, resulting in the loss of vital information and disrupting the overall semantic expression. On the other hand, textual semantic impairments introduce significant ambiguity into the original textual content, leading to potential misunderstanding for the recipient.

6.2.2.2 Robust Semantic Coding for Speech

The Ross-S2T is designed for communication scenarios where the transmitter and receiver users are across different linguistic backgrounds, enabling seamless speech-to-text conversion. The architecture of the robust semantic communication model for S2TT is shown in Figure 6.11. From the figure, when the input is clear speech, the deep semantic encoder compresses the speech sequence, $\boldsymbol{S}$, and extracts deep semantic features, $\boldsymbol{F}$. It is important to note that the deep semantic encoder directly learns the textual semantics in the target language. $\boldsymbol{F}$ is then mapped to symbols $\boldsymbol{X}$ via the NN-based channel encoder before transmission over the wireless channel.

The input of Ross-S2T extends to the corrupted speech, where some semantic information is lost due to semantic impairments. As shown in Figure 6.11, the corrupted speech, $\overline{\boldsymbol{S}}$, retains only limited semantic content. Specifically, the deep semantic compensator takes $\overline{\boldsymbol{S}}$ as input and produces the compensated deep

Figure 6.10 An example of different speech semantic impairments: (a) acoustic semantic impairments and (b) textual semantic impairments.

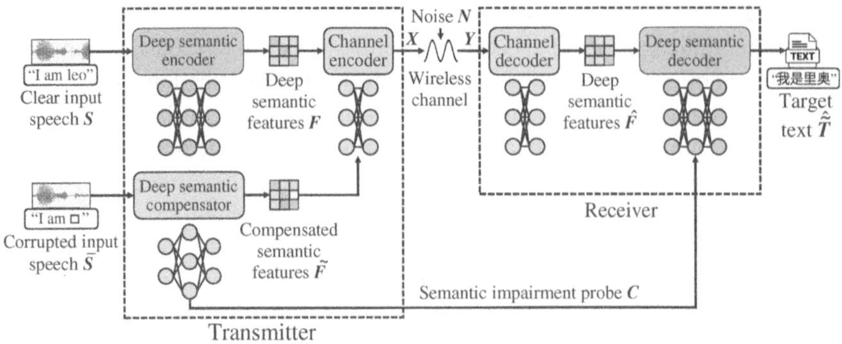

Figure 6.11 The model structure of robust semantic communications for S2TT.

semantic features, $\widetilde{F}$, predicting the impaired information and extracting textual semantic features in the target language simultaneously, formulated as

$$\widetilde{F} = \mathfrak{T}_{SC}(\overline{S}) \quad \text{w.r.t.} \quad \gamma, \tag{6.4}$$

Additionally, a semantic impairment probe, C, consisting of an index vector with the positions of the corrupted semantic features, is obtained and transmitted to the recipient via reliable channels. The goal of this probe is to enhance the semantic fidelity of the recovered deep semantic features by mitigating the ambiguity introduced by the corrupted information. Notably, the channel encoder, channel decoder, and deep semantic decoder, trained with clear speech input, are reused in scenarios involving the corrupted input speech.

6.2.2.3 Dual-compensator Mechanism

A dual-compensator mechanism consisting of a GAN-enabled deep semantic compensator at the transmitter and a probe-assisted compensator at the receiver is introduced, as shown in Figure 6.12.

6.2.2.3.1 Deep Semantic Compensator

The deep semantic compensator is responsible for estimating the corrupted semantic information in $\overline{S}$, extracting deep semantic features that closely resemble F, and returning the semantic impairment probe vector to capture the positional information of the corrupted deep semantic features. Specifically, the trained deep semantic encoder is used to obtain F as the real data for the discriminator. The deep semantic compensator, that is, the GAN generator, processes the corrupted speech, $\overline{S}$, and generates the fake data, $\widetilde{F}$, to deceive the discriminator by utilizing a one-dimensional (1D) convolutional neural network (CNN) module followed by

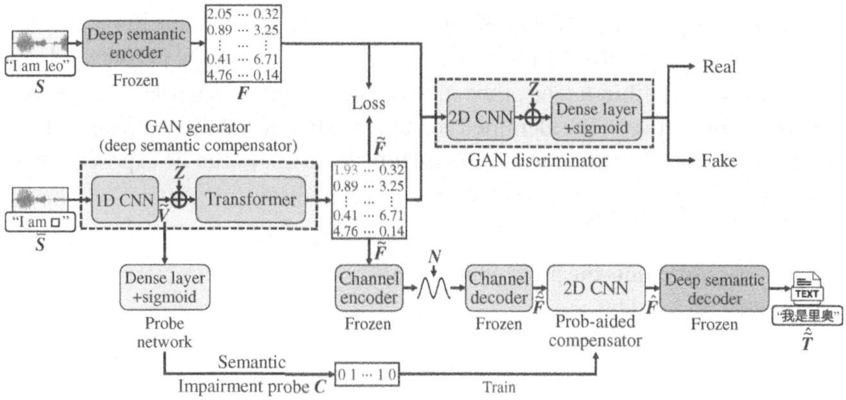

Figure 6.12 The dual compensator mechanism of Ross-S2T.

the latent space, $\boldsymbol{Z}$, and multiple transformer modules. Note that the intermediate semantic representation, $\widetilde{\boldsymbol{V}}$, is obtained as the output of the 1D CNN module. The discriminator distinguishes between real and fake data, incorporating a two-dimensional (2D) CNN module before the latent space $\boldsymbol{Z}$, followed by dense and sigmoid layers. The trainable NN parameters of the discriminator and generator are denoted as γ_{D} and γ_{G}, respectively. The loss function used for training the discriminator can then be expressed as:

$$\mathcal{L}_{\mathrm{D}}(\boldsymbol{S},\ \overline{\boldsymbol{S}};\ \gamma_{\mathrm{D}}) = \frac{1}{2}(\mathfrak{T}_{\mathrm{D}}(\mathfrak{T}_{\mathrm{S}}(\boldsymbol{S})) - 1)^2 + \frac{1}{2}\left(\mathfrak{T}_{\mathrm{D}}(\mathfrak{T}_{\mathrm{G}}(\overline{\boldsymbol{S}}))\right)^2, \tag{6.5}$$

where $\mathfrak{T}_{\mathrm{D}}(\cdot)$ and $\mathfrak{T}_{\mathrm{G}}(\cdot)$ are the discriminator and the generator, respectively.

Additionally, γ_{G} can be updated by the loss function as follows:

$$\begin{aligned}\mathcal{L}_{\mathrm{G}}(\boldsymbol{S},\ \overline{\boldsymbol{S}};\ \gamma_{\mathrm{G}}) &= \frac{1}{2}\xi\mathcal{L}_{\mathrm{MSE}} + \frac{1}{2}\mathcal{L}_{\mathrm{ADV}} \\ &= \frac{1}{2}\xi(\boldsymbol{F}-\widetilde{\boldsymbol{F}})^2 + \frac{1}{2}\left(\mathfrak{T}_{\mathrm{D}}(\mathfrak{T}_{\mathrm{G}}(\overline{\boldsymbol{S}})) - 1\right)^2,\end{aligned} \tag{6.6}$$

where ξ is a hyperparameter to scale the mean-squared error (MSE) loss, $\mathcal{L}_{\mathrm{MSE}}$.

The trained generator is deployed at the transmitter at the testing stage, creating realistic data to deceive the discriminator and producing compensated deep semantic features, $\widetilde{\boldsymbol{F}}$, with the least dissimilarity to $\boldsymbol{F}$.

6.2.2.3.2 *Probed-aided Compensator*

At the transmitter, a probe network is designed to generate the semantic impairment probe, $\boldsymbol{C}$, expressed as:

$$\boldsymbol{C} = \mathfrak{T}_{\mathrm{PN}}(\widetilde{\boldsymbol{V}}) \quad \text{w.r.t.} \quad \delta, \tag{6.7}$$

where $\mathfrak{T}_{\mathrm{PN}}(\cdot)$ refers to the probe network and δ is its trainable NN parameters.

To further improve the fidelity of the received deep semantic features, $\hat{\hat{\boldsymbol{F}}}$, the learned semantic impairment probe, $\boldsymbol{C}$, is employed to detect the corrupted deep semantic features within $\hat{\hat{\boldsymbol{F}}}$. The probe-aided compensator then works to minimize semantic errors between the identified corrupted features and their corresponding accurate features. The calibrated deep semantic features, $\hat{\boldsymbol{F}}$, can be expressed as:

$$\hat{\boldsymbol{F}} = \mathfrak{T}_{\text{PC}}(\hat{\hat{\boldsymbol{F}}}, \boldsymbol{C}) \quad \text{w.r.t.} \quad \zeta, \tag{6.8}$$

where $\mathfrak{T}_{\text{PC}}(\cdot)$ represents the probe-aided compensator, and ζ is the NN parameters, and it can be updated by

$$\mathcal{L}_{\text{PC}}(\widetilde{\boldsymbol{T}}, \hat{\hat{\boldsymbol{T}}}; \zeta) = -\sum_{l=1}^{L} p(\tilde{t}_l) \log p(\hat{t}_l), \tag{6.9}$$

where L indicates the number of tokens in $\widetilde{\boldsymbol{T}}$.

Through the compensation operation at the receiver, the understandability of the translated text, $\hat{\hat{\boldsymbol{T}}}$, is enhanced compared to scenarios where the received features are directly input into the deep semantic decoder.

6.2.2.4 Performance Metrics

The semantic textual similarity (STS) [13] is an efficient performance metric to evaluate the semantic fidelity of the translated text by comparing it to the accurate translation. To verify the superiority of Ross-S2T, a benchmark is provided by feeding the corrupted speech into the deep semantic encoder instead of the deep semantic compensator, named DeepSC-S2T. Another benchmark is considered, as shown in Figure 6.13, in which the SE module, MP-SENet [14], produces the enhanced speech before feeding it into the deep semantic encoder.

The STS results of different methods under acoustic semantic impairments and textual semantic impairments are presented in Figures 6.14 and 6.15, respectively, where the ground truth is obtained by directly passing the clear through the speech recognition module, conformer [15], and the machine translation pipeline, BART [16]. From the figures, the Ross-S2T attains higher BLEU and STS scores than the

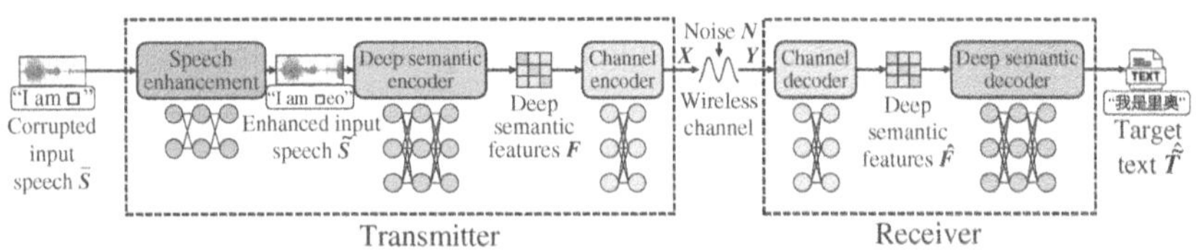

Figure 6.13 The model structure of benchmark.

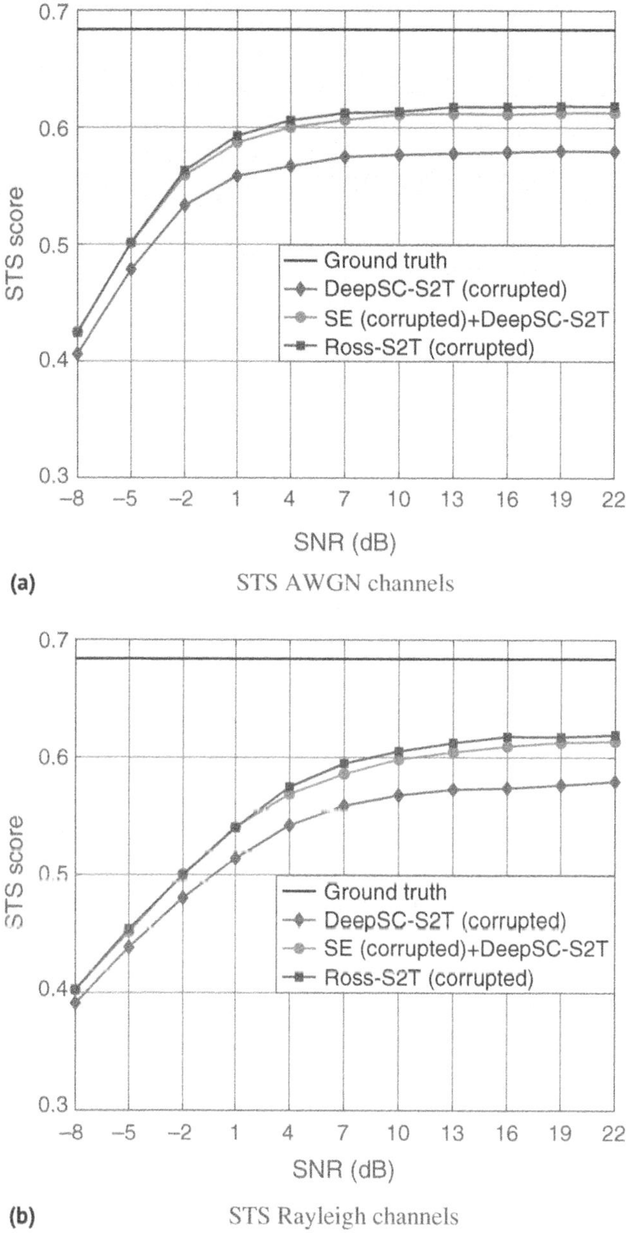

Figure 6.14 STS versus SNR for Ross-S2T on acoustic semantic impairments: (a) STS AWGN channels and (b) STS Rayleigh channels.

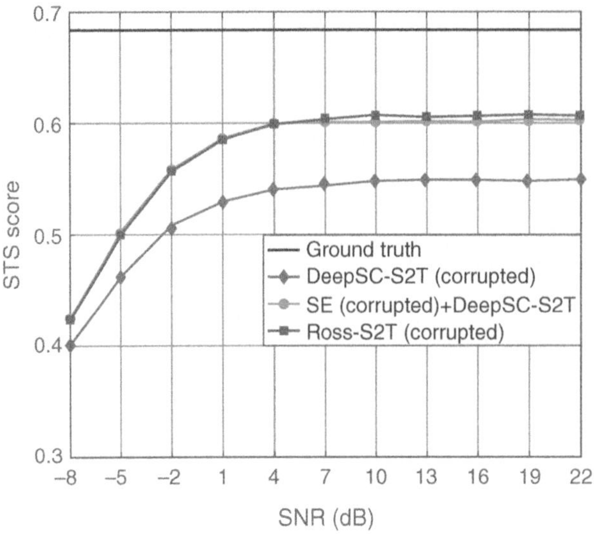

(a) STS AWGN channels

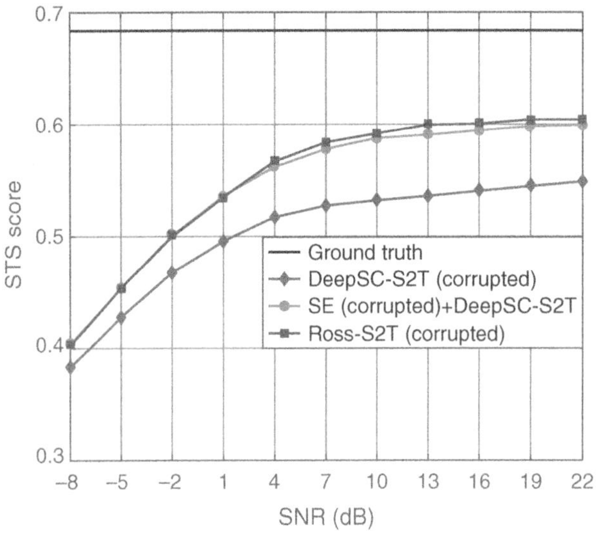

(b) STS Rayleigh channels

Figure 6.15 STS versus SNR for Ross-S2T on textual semantic impairments: (a) STS AWGN channels and (b) STS Rayleigh channels.

DeepSC-S2T under both AWGN and Rayleigh channels across all tested SNR values. It shows a comparable performance compared to the SE+DeepSC-S2T. Additionally, the acquired STS for Ross-S2T on textual semantic impairments is lower than the acoustic semantic impairments scenario, revealing the greater difficulty of recovering the original textual semantics within the speech from ambiguous oral expression than from noisy environments. Nevertheless, when the SNR drops below a certain threshold, the translated text becomes unintelligible in both scenarios, despite the marginally higher STS scores observed in the acoustic semantic impairments experiment.

6.2.3 Robust Semantic Communications for Image

In the field of image processing, the improvement of robustness mainly relies on adversarial purification methods. This method can eliminate disturbances in images, thereby obtaining clear and undamaged images. Current research on adversarial purification can be broadly categorized into three types.

The first type of adversarial purification method is based on image denoising. For instance, Xie et al. eliminate adversarial perturbations using feature denoising techniques, which not only demonstrates excellent robustness in white-box attacks but also effectively eliminates black-box attacks targeted at images [17]. Liao et al. proposed an image denoiser based on high-level representations, which trains the purification model by using the feature differences between clean and corrupted images as a loss function, effectively overcoming the error amplification issue of traditional denoisers [18]. Moreover, this method, with minimal sample training, can adapt to other images and unseen categories, demonstrating excellent generalization.

The second type of adversarial purification method involves using generative models to eliminate adversarial perturbations in images. Jia et al. introduced an adversarial sample defense method based on an E2E image generation model [19]. This method cleverly utilizes CNNs as the backbone of the generative model to maintain image structural information while removing adversarial perturbations, significantly enhancing the robustness of downstream classifiers. Samangouei et al. defended against adversarial attacks using the capability of generative adversarial models, retrieving images during testing that do not contain adversarial changes as approximate outputs of the model, effectively enhancing semantic fidelity in images [20]. Shi et al. proposed an online defense method combining self-supervised learning, which exhibits lower training complexity and stronger robustness, being the first to apply self-supervised learning to adversarial purification [21]. Hill et al. introduced an adversarial purification method based on energy models and Monte Carlo sampling, protecting naturally trained classifiers from adversarial attacks and serving as the first post-training defense strategy that does not require modifying the classifier training process [22].

The third type of adversarial purification method involves random perturbations to enhance the performance of adversarial purification. Yoon et al. proposed an adversarial purification method based on denoising score matching, which employs a strategy of injecting random noise to quickly eliminate adversarial perturbations in images, demonstrating excellent robustness against various forms of adversarial perturbations in practical tests [23]. Yang et al. disrupts adversarial noise structure by randomly dropping image pixels and reconstructing these pixels using matrix estimation while also strengthening the global structure of the original image, thereby improving the model's ability to eliminate inhibiting adversarial perturbations [24].

Although adversarial purification methods enhance the robustness of image semantics to a certain extent, they struggle to meet the demands of communication. Firstly, these methods often fail to fully consider channel constraints and prerequisites during design and implementation, making them exist as independent modules in communication systems, challenging their tight integration and optimization within the entire communication system.

Secondly, in the context of semantic communications, where the system transmits semantics, it is crucial to explore how to utilize semantic information for adversarial perturbation elimination and image reconstruction. This approach not only reduces the additional computational complexity resulting from repeated semantic extraction and interpretation but also effectively enhances the E2E semantic fidelity of semantic communication systems, providing users with more accurate and efficient information transmission experiences.

Three main aspects of work have been conducted. Firstly, a dataset containing different semantic damage intensities is constructed. Secondly, a dual-branch visual transformer is designed to extract multi-scale semantic information. Additionally, a feature fusion module based on the cross-attention mechanism is created to robustly build image semantics through the fusion of multi-scale semantic information. The working principle of the semantic encoder will be detailed next.

The semantic encoder consists of two branches. One branch is a fine-grained semantic extractor based on the visual transformer with some design improvements. In this branch, a semantic importance measurement module is built to gauge the importance of different semantic features. Subsequently, a dynamic semantic mask is constructed based on the obtained semantic importance to ensure that the model's self-attention mechanism focuses on crucial semantic information. The architecture of the fine-grained semantic extractor is illustrated in Figure 6.16.

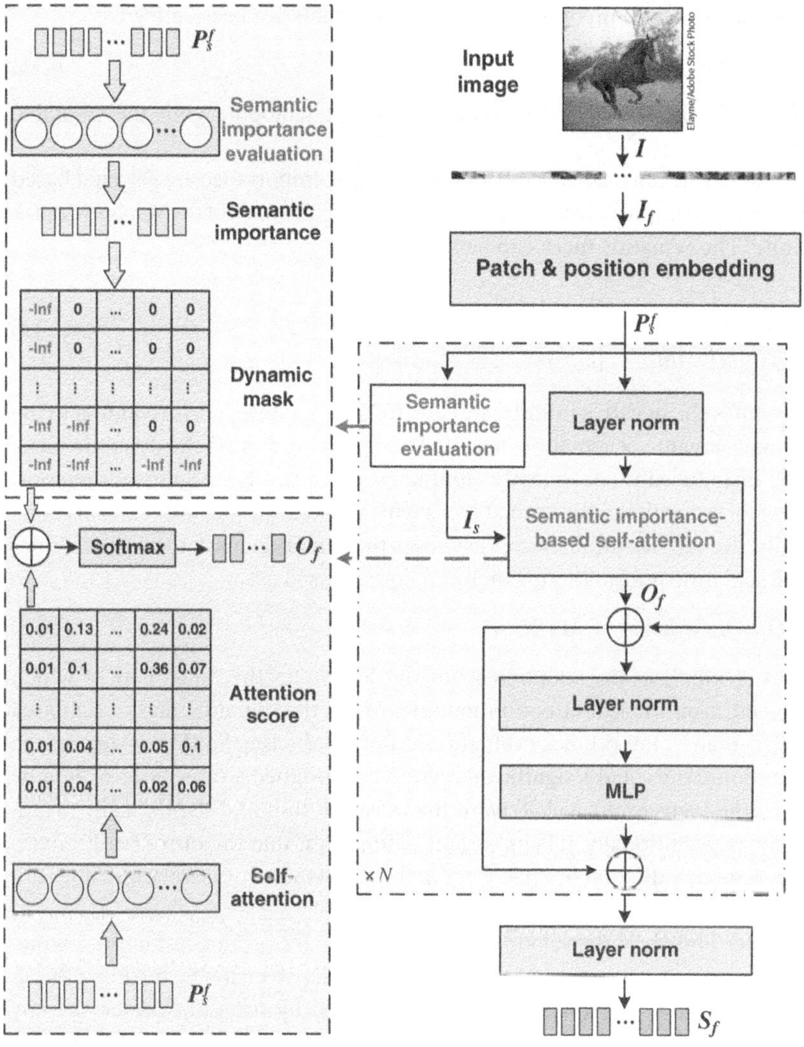

Figure 6.16 Architecture of the fine-grained semantic extractor. Wildestanimal/Alamy Stock Photo.

The specific process involves four steps. Firstly, the input semantic features are obtained, which can be represented as:

$$\boldsymbol{P}_s^f = f_\psi(\boldsymbol{I}_f), \tag{6.10}$$

where $f_\psi(\cdot)$ is the patch and position embedding module with the trainable parameter set $\boldsymbol{\psi}$.

Then, the semantic importance is measured, which can be described as:

$$\boldsymbol{I}_s = f_\varepsilon(\boldsymbol{P}_s^f), \tag{6.11}$$

where $f_\varepsilon(\cdot)$ is the semantic importance evaluation module with the trainable parameter set $\boldsymbol{\varepsilon}$.

The k semantic regions with the lowest semantic importance are selected based on the semantic importance, and the corresponding position of the semantic mask is infinite. The semantic mask can be represented as:

$$m_{i,j} = \begin{cases} 0, & \{i,\quad j\} \notin \mathbb{C}, \\ -\text{Inf}, & \{i,\quad j\} \in \mathbb{C}. \end{cases} \tag{6.12}$$

where $-\text{Inf}$ is the negative infinity, and $\mathbb{C}$ is the set of indexes corresponding to the top k smallest value of semantic importance. By incorporating the dynamic mask $\boldsymbol{M}_d$, the introduced model adeptly eliminates semantically insignificant regions, allowing for prioritizing the crucial areas with substantial semantic information.

Finally, the self-attention mechanism uses the semantic mask to construct fine-grained semantic information, which can be expressed as:

$$\boldsymbol{O}_f = \text{SoftMax}(\boldsymbol{A}_t + \boldsymbol{M}_d)\boldsymbol{V}, \tag{6.13}$$

where $\boldsymbol{A}_t$ symbolizes the attention score and $\boldsymbol{V}$ denotes the value, both of which are derived from the self-attention module. $\boldsymbol{M}_d$ is the dynamic mask originated through semantic importance evaluation. After conducting SoftMax($\cdot$) operation, the attention scores of less significant regions are assigned a value of zero, thereby directing the focus exclusively toward the most relevant and semantically meaningful areas. Additionally, it is important to highlight that the introduced semantic importance evaluation module is trained with the whole network and does not rely on supplementary labels.

The other branch of the semantic encoder is a coarse-grained semantic extractor that mainly constructs a hierarchical semantic extractor to further process the coarse-grained semantic information extracted by the transformer, thereby enhancing the model's ability to extract image semantic information. Figure 6.17 depicts the coarse-grained semantic extractor.

It is mainly divided into three steps. Firstly, semantic features extracted from the transformer are further processed through adaptive feature reconstruction, which can be expressed as:

$$S_j^i = \frac{\sum_{k \in \boldsymbol{R}} S_v^k}{|R|}, \tag{6.14}$$

where $\boldsymbol{R}$ is the set of level-i pooling regions and S_v^k is k-th value of the $\boldsymbol{S}_v$ in $\boldsymbol{R}$. A 3-level pooling is implemented for feature refinement. The first level of pooling

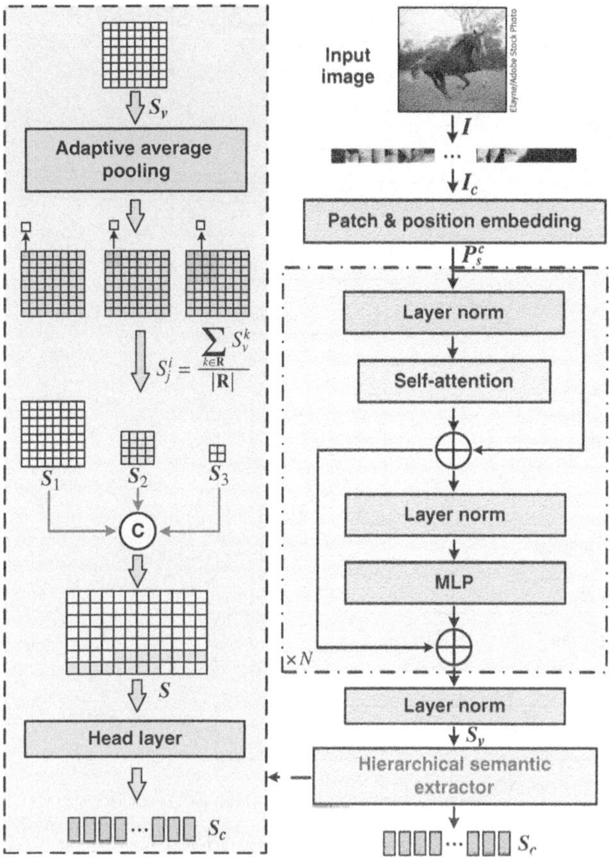

Figure 6.17 Architecture of the coarse-grained semantic extractor. Wildestanimal/Alamy Stock Photo.

preserves the dimensions of the feature map. In the second level, a 2×2 pooling region is applied, effectively halving the size of the feature map. Subsequently, the third level of pooling further reduces its size by half.

Then, these extracted features are merged, which is represented as:

$$\boldsymbol{S} = \text{concat}(\boldsymbol{S}^1, \boldsymbol{S}^2, \boldsymbol{S}^3), \tag{6.15}$$

where $\boldsymbol{S}^1$, $\boldsymbol{S}^2$, $\boldsymbol{S}^3$ is the semantic information extracted from the pooling process, while concat$(\cdot)$ is the concatenate operation, which is marked as in Figure 6.17.

Finally, these features are further encoded, which is represented as:

$$\boldsymbol{S}_c = f_v(\boldsymbol{S}), \tag{6.16}$$

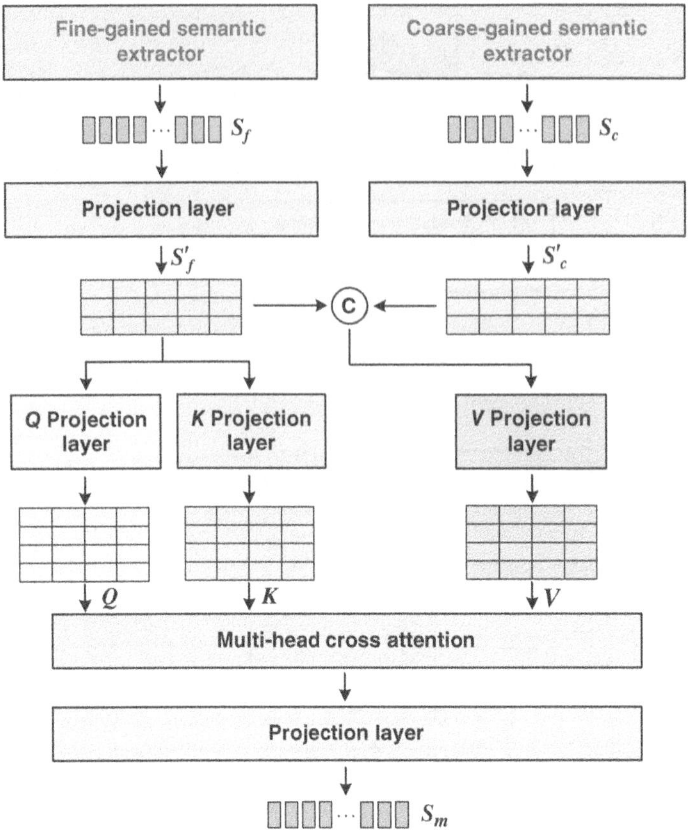

Figure 6.18 Architecture of the semantic fusion module.

where $f_{\nu}(\cdot)$ is the head layer with the trainable parameter set $\boldsymbol{\nu}$, which is composed of a linear layer and a sigmoid activation function.

Furthermore, a multi-granularity semantic information fusion module was designed to handle extracted coarse-grained and fine-grained semantic information. It utilizes cross-attention mechanisms to fuse semantic features of different granularities in three steps. Figure 6.18 illustrates the structure of the semantic fusion module.

Firstly, feature coding is carried out for coarse-grained and fine-grained semantic information. The processed coarse-grained semantics, $\boldsymbol{S}'_c$, can be represented as:

$$\boldsymbol{S}'_c = f_{\vartheta}(\boldsymbol{S}_c), \tag{6.17}$$

where $f_{\vartheta}(\cdot)$ is the coarse-grained projection layer with the trainable parameter set ϑ.

The fine-grained semantics, S_f', can be

$$S_f' = f_\theta(S_f), \tag{6.18}$$

where $f_\theta(\cdot)$ is the fine-grained projection layer with the trainable parameter set θ.

Secondly, the semantic association between coarse-grained semantic information and fine-grained semantic information is constructed by the cross-attention mechanism. The query, Q, the key, K, and the value, V, can be formulated as

$$Q = f_\kappa(S_f'), \tag{6.19}$$

$$K = f_\xi(S_f'), \tag{6.20}$$

$$V = f_\varpi(\text{concat}(S_f', S_c')), \tag{6.21}$$

where $f_\kappa(\cdot), f_\xi(\cdot), f_\varpi(\cdot)$ represents the Q projection layer with the trainable parameter set κ, the K projection layer with the trainable parameter set ξ, and the V projection layer with the trainable parameter set ϖ.

Finally, semantic fusion is carried out based on the results of semantic fusion. The multi-grained semantics is obtained by

$$S_m = f_\rho(\text{softmax}(Q \cdot K^T) \cdot V), \tag{6.22}$$

where $f_\rho(\cdot)$ is the projection layer with the trainable parameter set ρ to generate the final semantics.

Additionally, a semantic decoder based on the self-attention mechanism is constructed. This decoder utilizes multi-scale semantic information extracted by the encoder for robust semantic decoding, thereby enhancing the system's inhibitory performance against semantic damage. The system's loss function is divided into two parts: the first part includes cross-entropy loss between input images and ground truth, and the second part involves mean squared error loss between transmitted and received signals. Furthermore, a metric is designed to measure the intensity of image semantic damage, which projects images into semantic space using deep models, calculates semantic similarity, and derives the intensity of image semantic damage.

The experimental results are divided into two parts. Firstly, validation is conducted under different SNR ratios using AWGN and Rician channels. Four comparative algorithms are provided, including visual transformer-based systems, U-Net semantic communication systems, vector quantized deep semantic communication (VQ-DeepSC) communication systems, and better portable graphics (BPG)+low-density parity-check codes (LDPC) communication systems. Three evaluation metrics are presented, including learned perceptual image patch similarity (LPIPS), peak signal-to-noise ratio (PSNR), and classification accuracy of received images in downstream task models. It can be observed that the introduced method outperforms the comparative algorithms across all three metrics under both AWGN and Rician channels. In particular, in terms of classification accuracy, the introduced method improves by at least 20% compared to traditional communication systems, demonstrating its effectiveness in suppressing semantic damage.

Secondly, testing is carried out under different semantic damage intensities. It can be observed that as semantic damage intensity increases, the performance of all methods declines to varying degrees under both AWGN and Rician channels. This demonstrates that semantic damage indeed impacts the performance of communication systems, emphasizing the necessity of system design targeted at semantic damage for achieving semantic fidelity transmission. Moreover, with increasing semantic damage intensity, the introduced method achieved the best semantic fidelity, indicating its stronger robustness.

Furthermore, experiments are conducted on semantic segmentation tasks. The introduced method shows at least a 35% improvement in the mean intersection over union (mIoU) metric compared to traditional communication systems. Under different semantic damage intensities, the introduced method outperforms a visual transformer-based semantic communication system by at least 8% in terms of mIoU. This highlights the excellent robustness of the introduced method in semantic segmentation tasks.

6.3 Knowledge Base Semantic Impairments Suppression

Knowledge base semantic impairments usually refer to errors in the understanding of semantic information by semantic communication systems due to the incompleteness of the knowledge base. The main source of knowledge base semantic impairments is typically due to malicious attacks, where attackers may inject malicious data or instructions into the SKBs, undermining the reliability of the SKBs and leading to data pollution, misinformation, or system crashes.

6.3.1 Robust SKB

Knowledge base semantic impairments are modeled and divided into sample-dependent and sample-independent knowledge base semantic impairments. Sample-dependent knowledge base semantic impairments denote perturbation generated for each specific input sample, for example, each image or each text, which is related to specific sample content. This type of perturbation is usually generated by attackers targeting specific data samples to deceive DL models, which are almost imperceptible to humans but can significantly impact the model's robustness.

Sample-independent knowledge base semantic impairments represent perturbation that affects all input samples in the same way. This perturbation does not depend on specific sample content but applies the same interference to all inputs. Sample-independent knowledge base semantic impairments do not require knowledge of specific transmitted content or channel state information.

To further enhance the robustness of DL systems, the academic community has creatively developed a cutting-edge method of adversarial training to address knowledge base semantic impairments.

The core of adversarial training lies in utilizing gradient information to construct adversarial perturbations closely related to the model and integrating these perturbations into the training data to train the model. Through this unique training approach, the model can more accurately identify and respond to various potential disturbances, thereby enhancing its robustness.

Adversarial training can be described by the following optimization framework:

$$\min \mathbb{E}\,(\mathbf{x},\mathbf{y}) \in \boldsymbol{D}\,[\max \boldsymbol{L}\,(\boldsymbol{\theta},\, \mathbf{x} + \mathbf{S}_{\text{distortion}}),\, \mathbf{y}], \tag{6.23}$$

where $\mathbf{x}$ is the input to the model, $\mathbf{y}$ is the corresponding label, $\boldsymbol{D}$ is the dataset used for training, and $\mathbf{S}_{\text{distortion}}$ represents perturbations that could interfere with the model's accurate predictions. $\boldsymbol{L}(\cdot)$ is the loss function used by the model.

The implementation of adversarial training involves two key stages. First, by analyzing the parameters and structure of the downstream task model, perturbations that maximize the loss function are sought. This step aims to simulate uncertainty in the real world and reveal potential weaknesses of the model. Subsequently, based on these carefully crafted perturbations, the downstream task model is trained. During this phase, the model can learn to make accurate predictions even in the presence of adversarial perturbations, thereby improving the model's robustness and adaptability.

Research in adversarial training can be traced back to 2015 when Goodfellow et al. first introduced the fast gradient sign method (FGSM) [25]. This method normalized gradients using the sign function, laying a solid foundation for the field of adversarial training. Miyato et al. later proposed the fast gradient method (FGM) [2], which utilized L2 normalization techniques to search more finely and construct adversarial perturbations. Aleksander et al. further introduced the projected gradient descent method, which generated more effective adversarial samples through multiple iterative searches and optimizations [26]. To improve computational efficiency, Shafahi et al. introduced the FreeAT method, which significantly accelerated the generation of adversarial perturbations through innovative gradient reuse techniques, enhancing training efficiency [27]. Zhang et al. proposed the you only propagate once (YOPO) method, which further improved the efficiency and quality of adversarial training using the maximum principle [28]. Srinivasan et al. relaxed adversarial samples onto the target class distribution's underlying manifold, providing a new perspective on enhancing

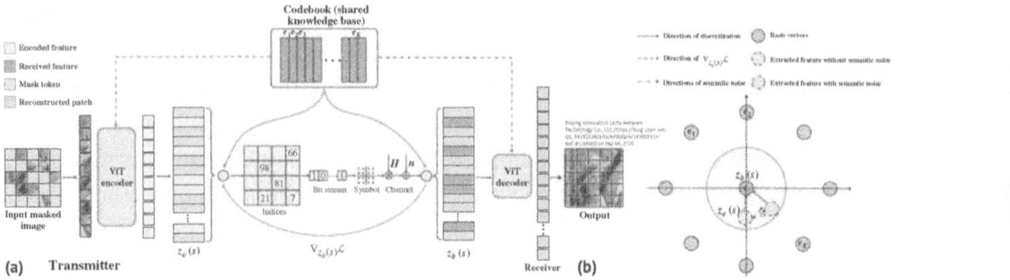

Figure 6.19 (a) The architecture of the masked VQ-VAE with discrete codebook; (b) The discrete codebook against knowledge semantic impairments.

the robustness of DL models [29]. These research achievements collectively drive the development of adversarial training and provide strong support for enhancing the robustness of DL systems.

Sample-dependent knowledge base semantic impairments are generated using the FGSM, while sample-independent knowledge base semantic impairments are generated based on the projected gradient descent method.

To build a robust semantic communication system, a masked vector quantized variational autoencoder (VQ-VAE) model is designed, and a novel masking strategy is developed. This model randomly masks parts of input images and attempts to reconstruct the missing blocks. The masked VQ-VAE is an autoencoder that can reconstruct the original image from partially observed data. It employs an asymmetric encoder-decoder architecture, where the encoder processes unmasked blocks only, while a lightweight decoder reconstructs the image from the encoded features and mask labels. The method is shown in Figure 6.19.

1) Discrete codebook and feature representation: A discrete codebook for compressing the semantic information is introduced, ignoring task-irrelevant features, noise, and imperceptible details. The codebook consists of a set of trainable base vectors. The encoder network outputs continuous encoding features, which are then mapped to the discrete indices of the base vectors in the trained codebook. This design simplifies codebook training and enhances the stability of the semantic communication system.

2) Feature importance module (FIM): To further enhance system robustness and transmission efficiency, an FIM is designed to dynamically learn and integrate the importance of features to train a DNN model that inherently suppresses task-irrelevant and noise-related features. The FIM extracts global feature information through global average pooling (GAP) and adjusts feature importance based on SNR, as shown in Figure 6.20.

The effectiveness of the introduced semantic communication system is validated through simulations. Figure 6.21 presents the classification accuracy of the presented masked VQ-VAE+FIM+autoencoder (AT) versus the reserving ratio.

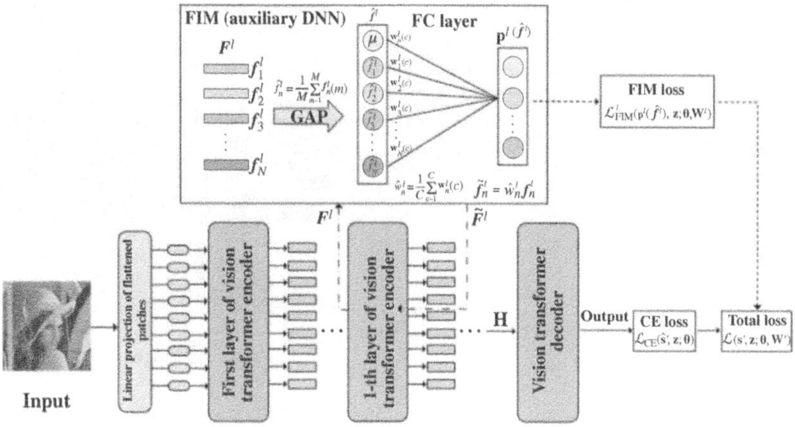

Figure 6.20 The architecture of the MAE with FIM.

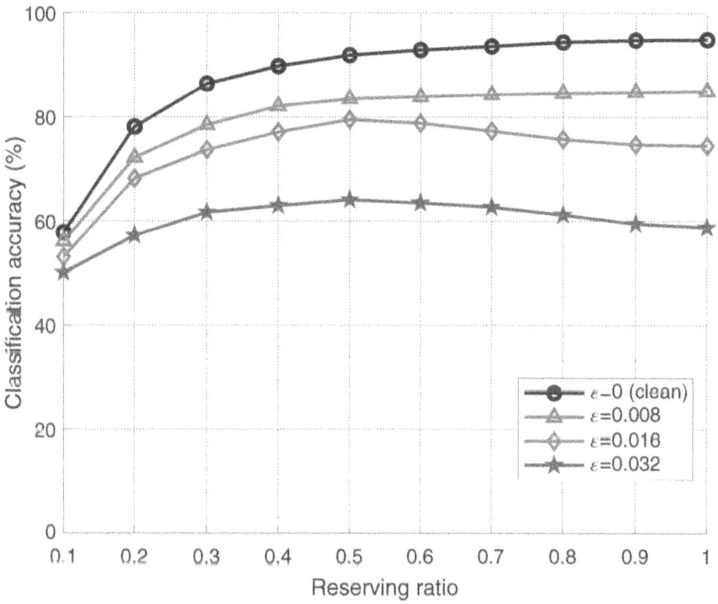

Figure 6.21 The classification accuracy of the presented scheme versus reserving ratio for different power of semantic impairments.

Figure 6.22 illustrates the recall@1 performance of the image retrieval task and the quality of image reconstruction under the power of semantic impairments. Under different SNRs, the introduced scheme demonstrates superior performance

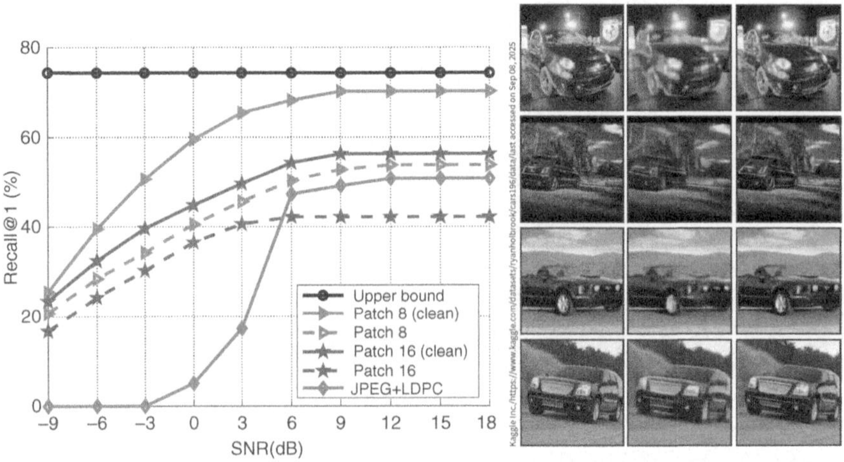

(a) The recall@1 performance of image retrieval. **(b)** The quality of image reconstruction.

Figure 6.22 The performance of image retrieval and reconstruction: (a) the recall@1 performance of image retrieval and (b) the quality of image reconstruction.

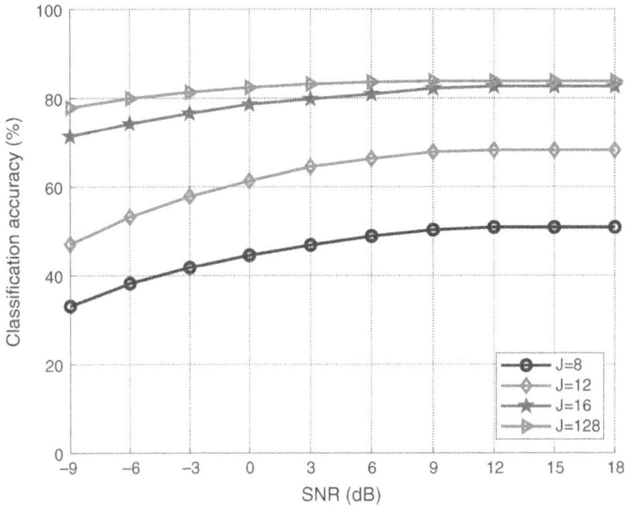

Figure 6.23 The classification accuracy of different codebook size J versus SNR.

in image classification, retrieval, and reconstruction tasks, particularly showcasing robustness in low SNR scenarios.

Additionally, the impact of different codebook sizes on classification accuracy and performance under various levels of semantic noise is investigated. Figure 6.23 presents the classification accuracy versus SNR for different codebook sizes. A larger codebook size has stronger robustness against semantic noise and more substantial representational capability for the whole dataset, which results in higher classification accuracy. The results indicate that the presented model effectively enhances the system's robustness to semantic noise while achieving high performance with low transmission overhead.

References

1 J. Bao, P. Basu, M. Dean, et al., "Towards a theory of semantic communication," in *IEEE Netw. Sci. Workshop*, West Point, NY, USA, 2011, pp. 110–117.

2 T. Miyato, A. M. Dai, and I. Goodfellow, "Adversarial training methods for semi-supervised text classification," in *Proc. Int. Conf. Learn. Represent. (ICLR)*, Toulon, France, 2017, pp. 1–11.

3 X. Peng, Z. Qin, X. Tao, et al., "A robust semantic text communication system," *IEEE Trans. Wireless Commun.*, vol. 23, no. 9, pp. 11372–11385, 2024.

4 Z. Zhao and H. Wang, "MaskGEC: improving neural grammatical error correction via dynamic masking," in *Proc. Assoc. Advancement Artif. Intell. (AAAI)*, vol. 34, New York, NY, USA, 2020, pp. 1226–1233.

5 W. Zhao, L. Wang, K. Shen, et al., "Improving grammatical error correction via pre-training a copy-augmented architecture with unlabeled data," in *Proc. N. Am. Chapter Assoc. Comput. Linguistics (NAACL)*, Minneapolis, MN, USA, 2019, pp. 156–165.

6 S. Zhang, H. Huang, J. Liu, and H. Li, "Spelling error correction with soft-masked BERT," in *Proc. Annual Meeting Assoc. Comput. Linguistics (ACL)*, Online, 2020, pp. 882–890.

7 S. Gandhi, P. Von Platen, and A. M. Rush, "ESB: a benchmark for multi-domain end-to-end speech recognition," *arXiv preprint arXiv:2210.13352*, 2022.

8 W. Li, H. Di, L. Wang, et al., "Boost transformer with BERT and copying mechanism for ASR error correction," in *Proc. Int. Joint Conf. on Neural Networks (IJCNN)*, online, 2021, pp. 1–6.

9 Y. Leng, X. Tan, L. Zhu, et al., "Fastcorrect: Fast error correction with edit alignment for automatic speech recognition," in *Proc. Neural Inform. Process. Systems (NeurIPS)*, vol. 34, online,2021, pp. 21708–21719.

10 F. Zhang, M. Tu, S. Liu, and J. Yan, "ASR error correction with dual-channel self-supervised learning," in *Proc. IEEE Int. Conf. Acoust. Speech Signal Process. (ICASSP)*, Singapore, 2022, pp. 7282–7286.

11 H. Xie, Z. Qin, G. Y. Li, and B.-H. Juang, "Deep learning enabled semantic communication systems," *IEEE Transactions on Signal Processing*, vol. 69, pp. 2663–2675, 2021.

12 Z. Weng, Z. Qin, and X. Tao, "Robust semantic communications for speech-to-text translation," *arXiv preprint arXiv:2403.05187*, 2024.

13 E. Agirre, C. Banea, C. Cardie, et al., "SemEval-2014 task 10: multilingual semantic textual similarity," in *Proc. Int. Workshop Semantic Eval. (SemEval)*, Dublin, Ireland, 2014, pp. 81–91.

14 Y.-X. Lu, Y. Ai, and Z.-H. Ling, "MP-SENet: a speech enhancement model with parallel denoising of magnitude and phase spectra," in *Proc. Annu. Conf. Int. Speech Commun. Assoc. (INTERSPEECH)*, Dublin, Ireland, 2023, pp. 3834–3838.

15 A. Gulati, J. Qin, C.-C. Chiu, et al., "Conformer: convolution-augmented transformer for speech recognition," in *Proc. Annu. Conf. Int. Speech Commun. Assoc. (INTERSPEECH)*, Shanghai, China, 2020, pp. 5036–5040.

16 M. Lewis, Y. Liu, N. Goyal, et al., "BART: denoising sequence-to-sequence pre-training for natural language generation, translation, and comprehension," in *Proc. Annu. Meet. Assoc. Comput Linguist. (ACL)*, online, 2020, pp. 7871–7880.

17 C. Xie, Y. Wu, L. v. d. Maaten, et al., "Feature denoising for improving adversarial robustness," in *Proc. IEEE Comput. Vis. and Pattern Recognit. (CVPR)*, 2019, pp. 501–509.

18 F. Liao, M. Liang, Y. Dong, T. Pang, X. Hu, and J. Zhu, "Defense against adversarial attacks using high-level representation guided denoiser," in *Proc. IEEE Comput. Vis. and Pattern Recognit. (CVPR)*, 2018, pp. 1778–1787.

19 X. Jia, X. Wei, X. Cao, and H. Foroosh, "Comdefend: an efficient image compression model to defend adversarial examples," in *Proc. IEEE Comput. Vis. and Pattern Recognit. (CVPR)*, 2019, pp. 6084–6092.

20 P. Samangouei, M. Kabkab, and R. Chellappa, "Defense-gan: protecting classifiers against adversarial attacks using generative models," in *Proc. Int. Conf. Learn. Represent. (ICLR)*, Vancouver, BC, Canada, 2018.

21 C. Shi, C. Holtz, and G. Mishne, "Online adversarial purification based on self-supervised learning," in *Proc. Int. Conf. Learn. Represent. (ICLR)*, online, 2020.

22 M. Hill, J. Mitchell, and S.-C. Zhu, "Stochastic security: adversarial defense using long-run dynamics of energy-based models," in *Proc. Int. Conf. Learn. Represent. (ICLR)*, online, 2021.

23 J. Yoon, S. J. Hwang, and J. Lee, "Adversarial purification with score-based generative models," in *Proc. Int. Conf. Machine Learn. (ICML)*, online, 2021, pp. 12062–12072.

24 Y. Yang, G. Zhang, D. Katabi, and Z. Xu, "ME-Net: towards effective adversarial robustness with matrix estimation," in *Proc. Int. Conf. Machine Learn. (ICML)*. California, CA, USA: PMLR, 2019, pp. 7025–7034.

25 I. J. Goodfellow, J. Shlens, and C. Szegedy, "Explaining and harnessing adversarial examples," in *Proc. Int. Conf. Learn. Represent. (ICLR)*, San Diego, CA, USA, 2015, pp. 1–11.

26 A. Madry, A. Makelov, L. Schmidt, et al., "Towards deep learning models resistant to adversarial attacks," in *Proc. Int. Conf. Learn. Represent. (ICLR)*, Vancouver, BC, Canada, 2018.

27 A. Shafahi, M. Najibi, A. Ghiasi, et al., "Adversarial training for free!" in *Proc. Neural Inform. Process. Systems (NeurIPS)*, Red Hook, NY, USA, 2019.

28 D. Zhang, T. Zhang, Y. Lu, et al., "You only propagate once: accelerating adversarial training via maximal principle," in *Proc. Neural Inform. Process. Systems (NeurIPS)*, vol. 32, Red Hook, NY, USA, 2019.

29 V. Srinivasan, C. Rohrer, A. Marban, et al., "Robustifying models against adversarial attacks by langevin dynamics," *Neural Networks*, vol. 137, pp. 1–17, 2021.

7

Generative AI-enabled Semantic Communications

This chapter explores the fusion of generative artificial intelligence (GAI) and SemCom. By combining cutting-edge generative models such as adversarial networks, diffusion models, and transformer-based frameworks with semantic transmission, the GAI-SemCom system accurately reconstructs intent even in noisy, bandwidth-limited environments while preserving the subtle context of multimodal data. The following sections introduce the framework and demonstration examples in which a GAI acts as both a semantic encoder and an imagination engine, able to reconstruct lost details and even generate context-aware content that is consistent with human cognition.

7.1 Introducing Generative Models to Semantic Communications

The rising adoption of artificial intelligence (AI) techniques in communications has led to the development of semantic-aware transmission strategies. These strategies prioritize transmitting the meaning or intent behind a message rather than just conveying raw bits. While many existing deep learning-based semantic communications focus on feature extraction, compression, and end-to-end optimization, the recent surge of GAI presents opportunities to enhance the quality, robustness, and flexibility of these systems. Generative models, including generative adversarial networks (GANs) [1], variational autoencoders (VAEs) [2], and diffusion-based models, have demonstrated remarkable capabilities in producing high-fidelity data samples that resemble real-world distributions, as shown in Figure 7.1. By applying these models to semantic communications, researchers aim to improve the efficiency and reliability of transmitted content and enable novel applications.

Deep Learning Enabled Semantic Communications, First Edition. Zhijin Qin et al.

Published 2026 by John Wiley & Sons, Inc.

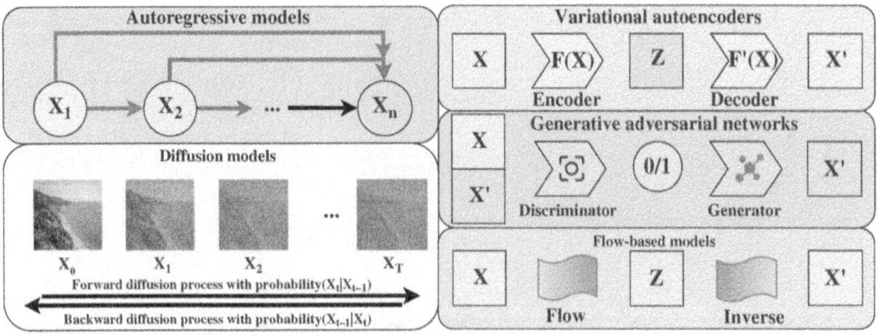

Figure 7.1 An illustration of typical GAI models. *Source*: Israel Ferrera/Unsplash/Public Domain.

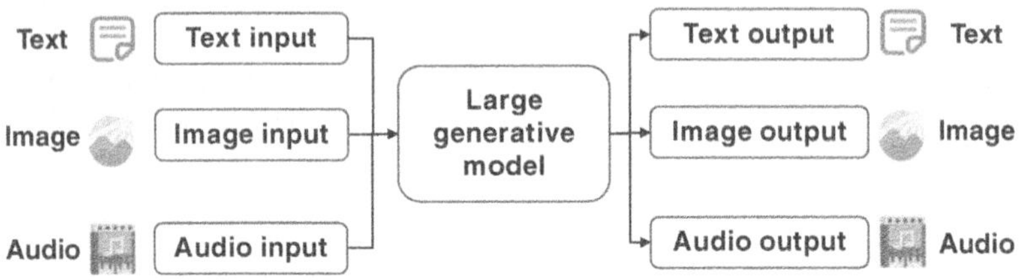

Figure 7.2 An illustration of multimodal large generative models.

A key advantage of generative models in semantic communications is their ability to learn comprehensive, high-dimensional representations of data distributions. Unlike models designed for classification or detection, which focus on encoding specific features, generative models aim to capture the underlying patterns that define a particular type of data, and enable cross-modal data transformation, as shown in Figure 7.2. This holistic understanding allows generative models to reconstruct or transform semantic content even when channel conditions degrade the transmitted signal. In semantic communication scenarios, disruptions such as noise, fading, and interference can distort critical semantic information. Generative models can restore semantic fidelity by recovering lost details or plausibly filling in missing content based on prior knowledge acquired during training. This approach surpasses traditional error-correcting codes and denoising algorithms by leveraging the context-specific semantics inherent in the data.

An illustrative example is the use of GANs, which involve a generator and a discriminator engaged in a minimax game. The generator produces data samples intended to deceive the discriminator, while the discriminator learns to distinguish between generated and authentic samples. Through this adversarial process, the generator effectively captures complex data distributions. In semantic communications, a well-trained generator can synthesize signal or message reconstructions that closely align with the original intent, even under bandwidth constraints that necessitate significant compression or in the presence of noise that corrupts portions of the signal. The discriminator acts as a supervisor, ensuring the fidelity of the reconstructed data to the semantic content. This adversarial training paradigm forms the basis for building robust semantic communication systems, offering resilience against channel impairments and enabling advanced features like style transformation or content augmentation.

Similarly, VAEs learn a lower-dimensional latent representation of data, which can be sampled to generate novel instances consistent with the original distribution. In semantic communications, a VAE-based transmitter can encode the essential semantics of a message into a compact latent space that is more resilient to channel noise. The receiver's decoder then reconstructs the essence of the data. By optimizing the variational encoder–decoder structure for end-to-end throughput, efficient semantic compression can be achieved without compromising the critical underlying meaning of the signal. Furthermore, manipulating the latent variables allows for interpretable modifications of semantic traits, enabling advanced features such as style adaptation or selective emphasis of specific semantic components.

Beyond GANs and VAEs, emerging generative paradigms like diffusion models [3], autoregressive models [4], and flow-based models [5] (as illustrated in Figure 7.1) offer additional benefits for semantic communications. These models enhance semantic communications' capabilities in various tasks, including image generation, text extraction, and handling complex data types. They enable semantic communications to improve overall communication quality and preserve subtle semantic elements under limited bandwidth and harsh channel conditions.

In conclusion, GAI provides a robust set of techniques for data representation, synthesis, and transformation that can significantly enhance semantic communication systems. Through generative modeling, semantic content can be preserved even in low-data-rate or noisy environments by leveraging learned distributions to fill in missing details or reconstruct entire signals. These capabilities exceed those of conventional encoding–decoding models in terms of adaptability and fidelity. As subsequent sections will demonstrate, integrating GAI paradigms into semantic communications not only offers incremental improvements but also paves the way for entirely new communication paradigms. In these paradigms, machines strategically leverage shared semantic knowledge to streamline information transfer and enrich user experiences [6].

7.2 Framework of Generative Semantic Communications

The framework of generative semantic communications, illustrated in Figure 7.3, introduces a revolutionary approach to information exchange by harnessing advanced AI-driven architectures. Central to this framework is the integration of

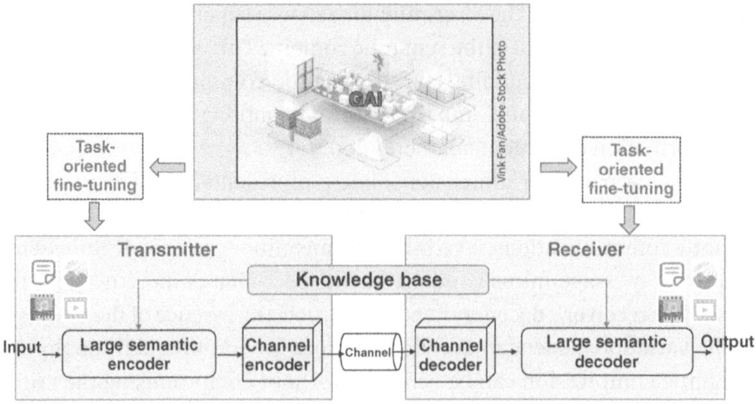

Figure 7.3 An illustration of large generative model-based semantic communication systems.

task-oriented fine-tuning and knowledge bases into a unified system that transcends conventional communication paradigms, facilitating the transmission of meaning rather than mere data symbols.

The process commences at the transmitter, where inputs – be they textual, visual, or multimodal – undergo processing through a semantic encoder. This encoder extracts and compresses high-level semantic features, a process enhanced by a dynamic knowledge base that provides contextual grounding and refines the representation of the input's intent. Once encoded, the compressed semantic information is transmitted through a channel optimized to minimize noise and maximize bandwidth efficiency [7].

At the receiver, a channel decoder reconstructs the transmitted features, which are then forwarded to a semantic decoder. This decoder generates outputs tailored to the recipient's context, ensuring not only accurate reconstruction but also meaningful interpretation. Task-oriented fine-tuning is critical throughout this architecture, refining both encoding and decoding processes to align with specific objectives and user requirements. Knowledge bases further enhance the system's capabilities by providing domain-specific insights and historical interaction data, promoting adaptability and precision. Collectively, these components form a robust framework that redefines communication as an exchange of actionable meaning. This approach paves the way for applications spanning intelligent customer service to cross-lingual and multimedia interactions.

In summary, this section presents a comprehensive framework for generative semantic communication systems, demonstrating how AI-driven architectures fundamentally transform information exchange. By integrating task-aware fine-tuning and knowledge grounding, these systems achieve unprecedented fidelity in reconstructing semantic intent across multimodal inputs. The outlined components – from adaptive encoding to context-aware generation – enable a paradigm shift from symbolic transmission to meaning-centric interaction. As subsequent validation cases will demonstrate, this framework not only enhances throughput efficiency but also opens new frontiers in human–machine collaboration by preserving nuanced semantics under constrained channels.

7.2.1 Main Components of Generative Semantic Communication System

This subsection details the components of the generative semantic communication system, beginning with the data preprocessing module, which aggregates and cleans raw data from diverse sources before extracting meaningful features using advanced techniques. Subsequently, GAI models are selected, trained, and fine-tuned to grasp domain-specific nuances and produce high-quality outputs. The semantic encoding layer maps the preprocessed data into a latent space, encoding it into a compact, semantic-rich representation for transmission. The semantic decoding layer decodes the received data, refines it, and reconstructs it into its original or similar form. Finally, the user interaction layer presents the decoded information to the user in an intuitive manner and incorporates feedback to enhance the system over time (Figure 7.4).

1) Data preprocessing module: This component forms the foundation of the system by collecting raw data from various sources, including text, audio, images, and video. The diversity of these sources ensures the system can handle a wide range of input types, making it adaptable for different applications [8]. Once acquired, the raw data undergoes normalization to ensure uniformity and

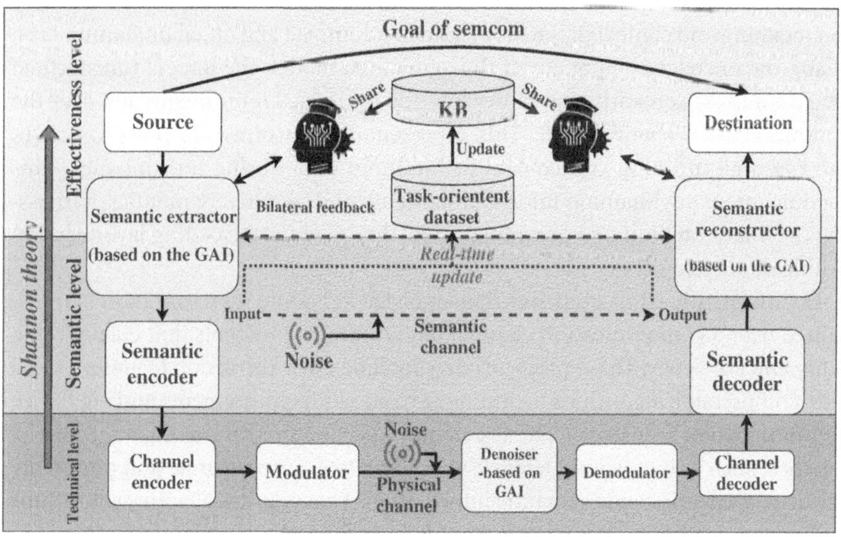

Figure 7.4 Three layers of the generative semantic communication system.

consistency. Cleaning processes are applied to remove noise, irrelevant information, and potential biases that could affect semantic content. Advanced techniques, such as natural language processing (NLP) for text, computer vision (CV) for images, or signal processing for audio, are leveraged to extract meaningful features from the preprocessed data. These features represent the essential aspects of the input data crucial for semantic communication [9]. As the entry point for system information processing, this module enables the semantic transformation of multimodal data. Its core value lies in converting the original signal into machine-understandable semantic primitives to mitigate noise and extract cross-modal semantic elements.

2) GAI models: Appropriate GAI models are selected based on the specific requirements of the data and communication context. These models may include VAEs, GANs, transformer-based models like generative pre-trained transformer (GPT-4) [10], or specialized models tailored for specific domains. The selected models undergo extensive training and fine-tuning on large and diverse datasets. This process helps the models understand domain nuances and generate high-quality outputs that accurately capture the semantics of the data. It also involves learning latent representations that efficiently encode the essential meaning and context of the input data. This module performs the core functions of semantic understanding and creation, enabling context-aware content generation. Technologies such as low-rank adaptation (LoRA) can optimize preferences for different objectives and meet the needs of task-oriented content, significantly enhancing the applicability of the generative semantic communication system.

3) Semantic encoding layer: The preprocessed data is mapped into a latent space where GAI models operate. This latent representation captures the essential meaning and context of the input data in a compact and efficient manner [11]. Using the encoder component of the generative model, the data is transformed into a compact, semantic-rich representation optimized for transmission over the communication channel [12]. This representation ensures the preservation of the essential meaning and context of the input data during transmission. This module integrates Shannon information theory and cognitive semantics, surpassing the Shannon limit and providing a new direction for the coding layer of communication systems.

4) Communication channel: The encoded semantic representation is transmitted over a communication channel using suitable protocols that ensure reliability and efficiency. These protocols may include error correction techniques and data compression algorithms to minimize bandwidth requirements and maximize communication robustness. The system must account for various channel characteristics, such as bandwidth, latency, and noise, to ensure accurate and minimally distorted transmission of semantic information. This requires adaptive algorithms and robust error-handling mechanisms to maintain the integrity of the transmitted data. As a bridge between the physical layer and the semantic layer, the focus of the algorithm should be on protecting semantics, which depends on the evaluation of semantic perception [13]. The adaptation of the channel and high alignment of semantic information are priorities for this module.

5) Semantic decoding layer: Upon receipt, the semantic representation is decoded using the decoder component of the generative model. This process reconstructs the data into its original or a similar meaningful form, ensuring the preservation of essential meaning and context. Depending on the specific application, additional refinement steps may be employed to enhance the quality of the reconstructed data. These steps could include post-processing algorithms or additional layers of generative models to improve the fidelity and accuracy of the reconstructed output.

6) User interaction layer: This layer is responsible for presenting the decoded information to the end-user in a user-friendly and intuitive manner. It may involve rendering text, audio, images, or interactive elements based on the user's preferences and context. The interface design should prioritize ease of use, accessibility, and engagement to ensure a positive user experience [14]. A feedback mechanism is integrated to gather user input and preferences, which can be used to improve the system's performance and user experience over time. This feedback loop enables the system to adapt to the user's needs and preferences, making it more personalized and responsive to user interactions. By continuously incorporating user feedback, the system can evolve and improve, ensuring it remains relevant and effective in a dynamic and evolving communication landscape.

7.2.2 Key Interactions and Processes

To comprehend the complex dynamics within a GAI system, it is essential to explore its key interactions and processes thoroughly. These components are crucial for ensuring the system's efficiency, adaptability, and user satisfaction.

1) Data-model interaction: The interaction between input data and GAI models is vital for maintaining high semantic fidelity. The models must continuously learn and adapt to evolving data patterns to remain relevant and accurate. This ongoing learning process involves feeding new data into the models, analyzing patterns, and adjusting model parameters accordingly. A two-way feedback reinforcement architecture is established, utilizing an incremental learning strategy to dynamically reconstruct the parameter space by continuously ingesting time series data streams [15]. This approach enables the system to better understand and generate content that aligns with the intended meaning and context, thereby enhancing overall performance.

2) Encoding–decoding cycle: The encoding and decoding processes form a critical closed loop within the system. Encoding involves converting input data into a format that the AI model can process, while decoding transforms the model's output back into a usable form. The effectiveness of semantic communications relies on the accuracy and efficiency of these operations. To ensure optimal performance, the system must continuously refine its encoding and decoding algorithms, improve semantic awareness measures and metrics, and leverage advanced techniques such as neural networks and NLP. By doing so, the system can minimize errors and maximize the fidelity of the transmitted information.

3) Adaptive communication: In dynamic environments, where communication channels and user requirements are constantly evolving, adaptability is essential. The system must be capable of making dynamic adjustments to transmission rates, error correction strategies, and data formats to maintain optimal performance. This involves real-time monitoring of communication conditions and user feedback, allowing the system to quickly adapt to changes and ensure seamless interaction [16]. By incorporating adaptive communication techniques, the system can better meet user expectations and deliver a more reliable and satisfying experience.

4) User experience enhancement: By leveraging the rich semantic information available within the system, GAI can tailor its output to better meet user needs. This involves understanding the context, intent, and preferences of the user and generating content that aligns with these factors. By doing so, the system can create more engaging and satisfying interactions, leading to higher user satisfaction and loyalty. To further enhance the user experience, the system can incorporate personalization techniques, such as recommending content based on user behavior and preferences, multi-dimensional characterization of users, and cross-domain migration of preference networks, to provide real-time feedback and support.

7.3 Demonstration of Generative Semantic Communication for Video Conferencing

Recent advances in GAI, especially in facial modeling and expression synthesis, have enabled the realization of semantic communication in practical applications. This demonstration introduces a GAI-powered semantic communication system for video conferencing that transmits talking-head videos with extremely low bitrates by encoding and transferring facial semantic information. At the receiver, realistic facial expressions are synthesized using generative models, achieving high-quality video reconstruction with bitrates as low as 21–30 kbps, while maintaining the speaker's identity and subtle facial expressions. The demonstration validates the effectiveness of integrating GAI with semantic communication, demonstrating its ability to enhance communication efficiency and maintain high subjective video quality in real-time wireless scenarios. The system not only reduces bandwidth consumption but also exhibits robust performance across

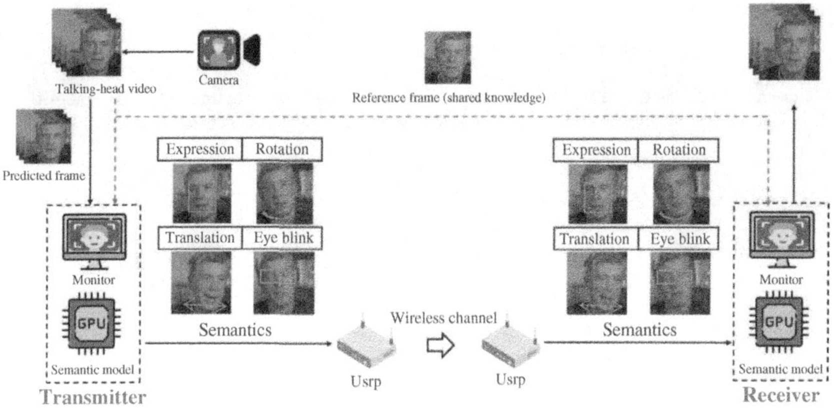

Figure 7.5 GAI-SemCom system for video conferencing. *Source*: Visual Geometry Group. / https://www.robots.ox.ac.uk/~vgg/data/voxceleb/ / last accessed on Sep 08, 2025.

different wireless environments, making it a promising solution for low-latency, high-efficiency communication in sixth generation (6G) and beyond.

The demonstration platform, as depicted in Figure 7.5, consists of a transmitter and a receiver connected through a wireless communication channel, with generative models driving the low-bitrate semantic communication. On the transmitter side, a camera captures the talking-head video, which is processed by a semantic encoder that extracts key facial features such as expressions, eye movements, and head rotations. These features are encoded into a compact semantic code, which is subsequently transmitted over the wireless channel via the universal software radio peripheral (USRP) B210 module. The receiver decodes this semantic bitstream and utilizes a semantic decoder powered by a GAI model to reconstruct the talking-head video. The reconstructed video, which accurately reflects the speaker's identity, facial expressions, and speaking style, is displayed on the monitor, ensuring real-time communication with minimal latency.

At the heart of this generative semantic communication system is its capability to model and compress talking-head videos by utilizing semantic key points derived from facial expressions and head movements. The system integrates a three-dimensional (3D) Morphable Model with advanced deep generative models to generate a compact, low-dimensional representation of facial expressions, head orientation, and other subtle motions. As illustrated in Figure 7.6, the semantic encoder captures facial features and motion-related information across multiple frames. These semantic features are then compressed into a bitstream and transmitted over the wireless channel.

To adapt to real-time wireless communication scenarios, the system dynamically adjusts to changing channel conditions and ensures consistent transmission quality even under varying signal-to-noise ratio (SNR) levels. The high efficiency

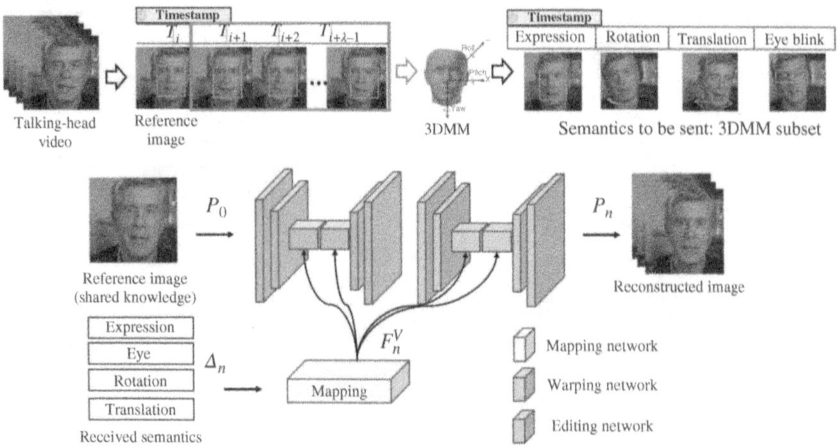

Figure 7.6 Semantic model for video conferencing.

of semantic encoding and the low-bit-rate transmission of semantic symbols significantly reduce network traffic, making the system suitable for bandwidth-constrained environments. Moreover, the incorporation of real-time GAI processing enables instantaneous reconstruction at the receiver, ensuring minimal latency and smooth communication.

In summary, this demonstration highlights the transformative potential of GAI-powered semantic communication in achieving low-bitrate, high-quality video conferencing. Beyond video calls, this technology holds promise for applications in telepresence, distance learning, augmented and virtual reality, and holographic conferencing, delivering immersive and efficient communication experiences. The integration of GAI also paves the way for multimodal systems that unify video, audio, and text into a cohesive semantic representation, significantly enhancing user experiences across diverse settings.

7.4 Applications of Semantic Communications in Other Scenarios

7.4.1 Immersive Communications

Unlike traditional communication paradigms, immersive communication scenarios prioritize the "meaning" of information over raw data. For instance, users in virtual environments require the perception of object positions, textures, and temperatures rather than pixel-level image transmission. Semantic communication extracts critical semantic features, such as object categories and spatial relationships, to significantly reduce data volume while enabling receivers to reconstruct complete scenes. Deep learning further enhances this framework by generating

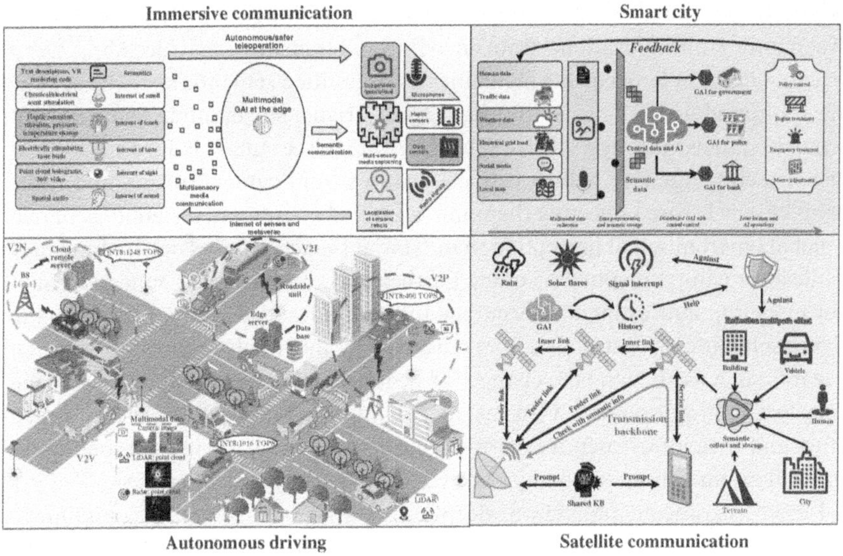

Figure 7.7 Four application scenarios of semantic communications.

realistic multimodal content, such as 3D models and haptic feedback, and predicting future scene dynamics based on semantic cues. In 6G-enabled internet of senses applications, which aim to deliver multi-sensory signals (visual, auditory, tactile, olfactory) for hyper-realistic virtual experiences, as shown in Figure 7.7. Studies show that combining semantic compression with deep learning reduces bandwidth requirements by up to 99.93%, while achieving sub-millisecond latency through end-to-end semantic synchronization [17].

For practical implementation, consider metaverse applications that require processing high-dimensional data, such as point clouds and mesh models. Traditional methods relying on raw data transmission suffer from high latency and resource inefficiency. Goal-oriented semantic communication addresses this by extracting semantic features, such as motion trajectories and material properties, and leveraging deep learning for dynamic scene reconstruction. For example, 360-degree videos captured by drones can be semantically encoded into textual descriptions, with receivers using deep learning models to produce WebXR code for rendering 3D objects and synchronizing haptic feedback on head-mounted displays [18].

7.4.2 Autonomous Driving

This advanced communication architecture significantly enhances vehicles' environmental understanding by transmitting the essential meaning of information rather than raw data. For instance, in cooperative lane-changing scenarios, the system conveys not only vehicle coordinate data but also semantic information, such as lane-change intention. The integration of deep learning strengthens every stage of semantic communication – from encoding and transmission to decoding. Deep learning maps heterogeneous modalities into a unified semantic space by acquiring multimodal data (e.g., traffic signs, pedestrians, surrounding vehicles, and acoustic signals) through diverse sensors. A notable example is Bird's-Eye View fusion technology [19], which extracts features from camera images and radar point clouds before projecting them into a unified coordinate system to generate a global environmental perception map. This approach reduces data redundancy while improving perceptual accuracy. Furthermore, deep learning enhances channel modeling and estimation accuracy through advanced techniques, enabling robust semantic reconstruction from noisy signals and timely response to potential hazards such as sudden lane changes or pedestrian crossings [20]. By leveraging deep learning, autonomous vehicles can communicate more effectively with each other and with infrastructure, sharing critical information about road conditions, traffic flow, and obstacles to improve overall road safety and efficiency.

Despite significant progress in deep learning enabled semantic communication technologies, several challenges remain. A primary concern is balancing the trade-off between analog transmission's fidelity advantages and digital systems' noise resilience, which can lead to suboptimal resource allocation where critical safety messages may be delayed while non-essential data consumes excessive bandwidth. Additionally, the isolated processing pipelines for perception, planning, and control tasks can result in redundant feature extraction and fragmented environmental understanding, increasing computational overhead and leading to inconsistent decision-making. Finally, the semantic mismatch between localized vehicle perceptions and global traffic management systems can cause delayed situational awareness updates and conflicting maneuver recommendations, reducing the effectiveness of cooperative driving and traffic flow optimization. Future research directions should focus on developing context-aware modulation schemes, creating unified semantic representation spaces, and constructing hierarchical knowledge graphs to address these challenges.

7.4.3 Smart Cities

Smart cities increasingly rely on deep learning to synthesize and interpret heterogeneous data streams, including traffic dynamics, energy consumption patterns, and public safety metrics [21]. By integrating deep learning with semantic communication frameworks, urban systems can distill complex data into actionable insights while minimizing transmission overhead. For instance, deep learning processes real-time traffic data from Internet-of-Things (IoT) sensors and connected vehicles to predict congestion hotspots and dynamically optimize signal timings. Semantic communication protocols ensure low-latency dissemination of traffic predictions to navigation platforms and municipal control centers, greatly reducing average commute times in pilot deployments [22]. This synergy enables adaptive routing recommendations that account for both macroscopic traffic flow and microscopic vehicle interactions.

Despite the advancements in deep learning enabled semantic communication for smart cities, several challenges remain. These include the need for efficient resource allocation and responsiveness at network edges, achieving cross-domain semantic interoperability across urban management systems, and ensuring security and robustness against evolving cyber threats. Future research should focus on developing adaptive model partitioning strategies, creating unified semantic mapping architectures, and integrating neuro-symbolic frameworks to address these challenges and enhance the effectiveness of smart city operations.

7.4.4 Satellite Networks

The integration of deep learning with semantic communications presents a revolutionary pathway for satellite communications [23]. By employing U-Net-like models to decompose images into multi-class semantic features, the system can transmit only essential information – such as vehicle contours or structural outlines – achieving significant reduction in bandwidth requirements compared to conventional methods. The semantic-aware channel encoder dynamically adapts its strategy based on channel conditions, prioritizing critical features during low SNR scenarios. This deep learning enabled semantic communication framework significantly enhances communication reliability through intelligent content protection mechanisms. When encountering sudden interference, the system leverages deep learning models and historical image correlations to maintain robust connectivity. Furthermore, replacing traditional cyclic redundancy check (CRC) checks with a two-tiered detection system – using a lightweight rough detector for initial semantic integrity assessment followed by precise verification at ground gateways – reduces unnecessary retransmissions by approximately 35% while improving overall transmission rates [24].

Despite the promising potential of deep learning enabled semantic communication in satellite networks, several challenges remain. These include addressing the heterogeneity in data types from satellite terminals, overcoming latency in low Earth orbit (LEO) satellite-ground links, and managing geographic semantic drift phenomena. Future research should focus on constructing cross-modal semantic distillation networks, deploying on-orbit semantic computing chiplets, and developing regional knowledge distillation architectures to enhance the efficiency and reliability of satellite communications.

References

1 I. Goodfellow, J. Pouget-Abadie, M. Mirza, et al., "Generative adversarial networks," *Commun. ACM*, vol. 63, no. 11, pp. 139–144, 2020.

2 D. P. Kingma and M. Welling, "Auto-encoding variational bayes," in *Proc. Int. Conf. Learn. Represent. (ICLR)*, 2014.

3 F.-A. Croitoru, V. Hondru, R. T. Ionescu, and M. Shah, "Diffusion models in vision: a survey," *IEEE Trans. Pattern Anal. Mach. Intell.*, vol. 45, no. 9, pp. 10850–10869, 2023.

4 E. Zhao, P. Awasthi, Z. Chen, et al., "Semantic routing via autoregressive modeling," *Adv. Neural Inf. Process. Syst.*, vol. 37, pp. 10060–10087, 2024.

5 G. Harshvardhan, M. K. Gourisaria, M. Pandey, and S. S. Rautaray, "A comprehensive survey and analysis of generative models in machine learning," *Comput. Sci. Rev.*, vol. 38, p. 100285, 2020.

6 H. Du, Z. Li, D. Niyato, et al., "Enabling ai-generated content (aigc) services in wireless edge networks," *arXiv preprint arXiv:2301.03220*, 2023.

7 R. Cheng, Y. Sun, D. Niyato, et al., "A wireless ai-generated content (AIGC) provisioning framework empowered by semantic communication," *IEEE Trans. Mob. Comput.*, 2024.

8 S. García, S. Ramírez-Gallego, J. Luengo, et al., "Big data preprocessing: methods and prospects," *Big Data Anal.*, vol. 1, pp. 1–22, 2016.

9 P. Mishra, A. Biancolillo, J. M. Roger, et al., "New data preprocessing trends based on ensemble of multiple preprocessing techniques," *TrAC Trends Anal. Chem.*, vol. 132, p. 116045, 2020.

10 A. Radford, K. Narasimhan, T. Salimans, and I. Sutskever, "Improving language understanding by generative pre-training," https://gwern.net/doc/www/s3-us-west-2.amazonaws.com/d73fdc5ffa8627bce44dcda2fc012da638ffb158, 2018.

11 X. Peng, Z. Qin, X. Tao, et al., "A robust semantic text communication system," *IEEE Trans. Wireless Commun.*, vol. 23, pp. 11372–11385, 2024.

12 E. Bourtsoulatze, D. B. Kurka, and D. Gündüz, "Deep joint source-channel coding for wireless image transmission," *IEEE Trans. Cogn. Commun. Netw.*, vol. 5, no. 3, pp. 567–579, 2019.

13 L. Yan, Z. Qin, R. Zhang, et al., "Resource allocation for text semantic communications," *IEEE Wireless Commun. Lett.*, vol. 11, no. 7, pp. 1394–1398, 2022.

14 J. Yeon, Y. Jung, Y. Baek, et al., "User preferences on a generative ai user interface through a choice experiment," *Int. J. Hum–Comput. Int.*, pp. 1–12, 2024.

15 H. Du, G. Liu, D. Niyato, et al., "Generative al-aided joint training-free secure semantic communications via multi-modal prompts," in *Proc. IEEE Int. Conf. Acoust., Speech, Signal Process. (ICASSP)*, 2024, pp. 12896–12900.

16 H. Gao, G. Yu, and Y. Cai, "Adaptive modulation and retransmission scheme for semantic communication systems," *IEEE Trans. Cogn. Commun. Netw.*, vol. 10, no. 1, pp. 150–163, 2024.
17 N. Sehad, L. Bariah, W. Hamidouche, et al., "Generative ai for immersive communication: the next frontier in internet-of-senses through 6g," *IEEE Commun. Mag.*, 2024.
18 Z. Wang, N. Li, Y. Deng, and A. H. Aghvami, "Goal-oriented semantic communications for metaverse construction via generative ai and optimal transport," *arXiv preprint arXiv:2411.16187*, 2024.
19 J. Lu, W. Yang, Z. Xiong, et al., "Generative artificial intelligence-enhanced multimodal semantic communication in internet of vehicles: system design and methodologies," *IEEE Veh. Technol. Mag.*, 2025.
20 S. Zhang, I.-L. Yen, F. Bastani, et al., "A semantic model for information sharing in autonomous vehicle systems," in *Proc. Int. Conf. Semantic Comput. (ICSC)*, 2017, pp. 32–39.
21 A. Von Richthofen, P. Herthogs, M. Kraft, and S. Cairns, "Semantic city planning systems (scps): a literature review," *J. Plan. Lit.*, vol. 37, no. 3, pp. 415–432, 2022.
22 K. P. Muriuki, J. O. Okello, and J. Chepkoech, "Advanced intelligent traffic management system (AITMS): a generative AI-enhanced model," in *IEEE PES/IAS PowerAfrica*, 2024, pp. 1–3.
23 F. Fourati and M.-S. Alouini, "Artificial intelligence for satellite communication: a review," *Intell. Converg. Netw.*, vol. 2, no. 3, pp. 213–243, 2021.
24 P. Jiang, C.-K. Wen, X. Li, et al., "Semantic satellite communications based on generative foundation model," *arXiv preprint arXiv:2404.11941*, 2024.

8

Conclusion and Challenges

The great potential of deep learning enabled semantic communications has been discussed in this book, including semantic communications for source reconstruction, task-oriented semantic communication systems, semantic impairment suppression mechanisms, joint sensing and semantic communication paradigms, and large model-driven semantic communications. However, some fundamental technical bottlenecks in semantic communications are still under investigation. For example, communications resources are reduced by significantly condensing the input source, while more computing resources are required to achieve a high compression ratio. As a result, the tradeoff between communications resources and computing resources needs to be resolved. Moreover, due to the nature of deep learning (DL) techniques, DL enabled semantic communications contend with limitations and constraints, such as time-consuming model pretraining, large data requirements, and poor mathematical interpretability.

To accelerate the deployment of speech semantic communications in practical communication networks, it is believed that the following challenges are of significant importance and need to be addressed in the future.

1) One model for all in semantic communications: The transmission goal of semantic communications is dynamic and depends on user requests. According to the latest literature review, diverse semantic frameworks are tailored to serve specific downstream intelligent tasks. However, it is impossible to deploy multiple semantic decoders in the user terminal to perform different tasks due to the limited computational resources of mobile devices. Inspired by this, a unified semantic communication paradigm is necessary to execute various intelligent tasks using a single powerful semantic decoder at the receiver. Moreover, the design of the semantic codecs and the channel codecs is associated with the channel environments, and the performance of semantic communication systems depends on the channel conditions in the training stage. Some existing research works

Deep Learning Enabled Semantic Communications, First Edition. Zhijin Qin et al.

Published 2026 by John Wiley & Sons, Inc.

developed semantic communication systems that obtain robustness to a specific channel type, such as single-input single-output (SISO) fading channels. Nevertheless, such semantic communication systems only maintain stability under a particular type and are considered insufficient due to the unpredictable and uncontrollable channel environments in practical communication scenarios. Therefore, it is crucial to design unified semantic communication systems that are robust to numerous intelligent tasks and practical transmission environments.

2) Semantic communications compatibility: In semantic communication systems, the constellations vary with different source data, aligning well with analogue communication paradigms. However, digital communication systems dominate practical applications, as existing transmission hardware is primarily designed for digital formats and lacks compatibility with analogue systems. Redesigning hardware to accommodate semantic communication systems would entail significant costs. Thus, ensuring compatibility between semantic and digital communication systems is a critical challenge for deploying semantic communication technologies in real-world transmission scenarios.

3) Efficient algorithms for the training and inference of large semantic models: One of the primary challenges in the field of generative semantic communications is the need for more efficient and scalable algorithms for the training and inference of large semantic models. Training these large semantic models requires significant computational resources and time, which can be a significant barrier to their widespread adoption. Furthermore, inference can also be computationally intensive, making it difficult to deploy these models in resource-constrained environments. To overcome this challenge, researchers are exploring various techniques, such as distributed training, model pruning, and quantization, to reduce the computational burden and improve scalability. Developing more efficient algorithms that can reduce the resource requirements for training and inference is crucial for enabling the widespread adoption of generative semantic communication systems.

4) Semantic communications standardization: The candidate key performance indicators (KPIs) for sixth generation (6G) and beyond are expected to include the following table. From the table, it is evident that 6G networks will require a significant improvement in terms of data rates, latency, reliability, and energy efficiency compared to fifth-generation (5G) networks. The key difference between 5G and 6G is artificial intelligence (AI). 6G designs for AI and also employs AI to improve the performance of the network, the aim of which is connectivity and beyond connectivity. For example, the agentic multimodality communication system is one of the user case, in which the communication carrier performs as the agent to understand the intention of the user and deliver the accurate intention to another intelligent service. In this context, the standardization of semantic communication technologies is around the meaning behind users' transmission and is suitable for 6G. However, there is a lack of consensus on the best practices for integrating semantic communication systems into future wireless networks. For example, how to comply with conventional communication systems with semantic communications. Therefore, the new protocols will be a critical challenge in realizing the full potential of semantic communications in 6G networks.

KPI	5G	6G
Traffic capacity	10 Mbps/m^2	1–10 Gbps/m^3
Data rate DL	20 Gbps	1 Tbps
Data rate UL	10 Gbps	1 Tbps
Uniform user experience	50 Mbps 2D everywhere	10 Gbps 3D everywhere
Mobility	500 km/h	1000 km/h
Latency (radio interface)	1 ms	0.1 ms
Jitter	NS	1 μs
Communication reliability	$1-10^{-5}$	$1-10^{-9}$
Inference reliability	NS	TBD
Energy/bit	NS	1 pJ/bit
Energy/goal	NS	TBD
Localization precision	10 cm on 2D	1 cm on 3D

2D, two-dimensional; 3D, three-dimensional.

5) Neuro-symbolic AI-enabled semantic communications: The explainability of data is another key challenge for semantic communication systems. The explainability of data is crucial for ensuring the trustworthiness of AI-enabled systems and for enabling users to understand the decisions made by these systems. However, the explainability of data is a challenging task, as AI models often operate as black boxes, making it difficult to understand how they arrive at their decisions. To address this challenge, researchers are exploring neuro-symbolic AI approaches that combine the strengths of neural networks and symbolic reasoning to create more interpretable and explainable AI models. By integrating neuro-symbolic AI techniques into semantic communication systems, researchers can develop more transparent and trustworthy systems that enable users to understand the decisions made by AI models. This will be crucial for ensuring the trustworthiness of semantic communication systems and for enabling users to understand the decisions made by these systems.

Index

Deep Learning Enabled Semantic Communications, First Edition. Zhijin Qin et al.

Published 2026 by John Wiley & Sons, Inc.